Olcay Akman, Christopher Hay-Jahans
Biomathematical Modeling

Also of Interest

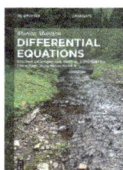

Differential Equations. Solving Ordinary and Partial Differential Equations with Mathematica®
Marian Mureşan, 2024
ISBN 978-3-11-141109-5, e-ISBN (PDF) 978-3-11-141139-2,
e-ISBN (EPUB) 978-3-11-141204-7

Probability Theory and Statistics with Real World Applications. Univariate and Multivariate Models Applications
Peter Zörnig, 2024
ISBN 978-3-11-133220-8, e-ISBN (PDF) 978-3-11-133227-7,
e-ISBN (EPUB) 978-3-11-133232-1

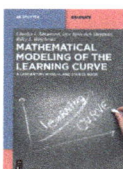

Mathematical Modeling of the Learning Curve. A Laboratory Manual and Source Book
Charles I. Abramson, Igor Igorevich Stepanov, Riley J. Wincheski, 2024
ISBN 978-3-11-131173-9, e-ISBN (PDF) 978-3-11-131367-2,
e-ISBN (EPUB) 978-3-11-131390-0

Bioinformatics. Drug Discovery
Anil K. Sharma, Varruchi Sharma (Eds.), 2024
ISBN 978-3-11-156788-4, e-ISBN (PDF) 978-3-11-156858-4,
e-ISBN (EPUB) 978-3-11-156882-9

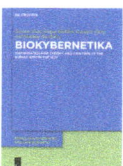

BIOKYBERNETIKA. Mathematics for Theory and Control in the Human and in Society
Jochen Mau, Sergey Mukhin, Guanyu Wang, Shuhua Xu (Eds.), 2024
ISBN 978-3-11-134197-2, e-ISBN (PDF) 978-3-11-134199-6,
e-ISBN (EPUB) 978-3-11-134204-7

Olcay Akman, Christopher Hay-Jahans

Biomathematical Modeling

Methods and Software

DE GRUYTER

Mathematics Subject Classification 2020
Primary: 92-04, 92-08, 92-10; Secondary: 34A30, 62H30

Authors

Prof. Olcay Akman
Illinois State University
Mathematics Department
Normal, IL 61790-4520
USA
oakman@ilstu.edu

Prof. Christopher Hay-Jahans
University of Alaska Southeast
Department of Natural Sciences
11066 Auke Lake Way
Juneau, AK 99801
USA
cnhayjahans@alaska.edu

ISBN 978-3-11-160952-2
e-ISBN (PDF) 978-3-11-160956-0
e-ISBN (EPUB) 978-3-11-160994-2

Library of Congress Control Number: 2025931501

Bibliographic information published by the Deutsche Nationalbibliothek
The Deutsche Nationalbibliothek lists this publication in the Deutsche Nationalbibliografie;
detailed bibliographic data are available on the Internet at http://dnb.dnb.de.

www.degruyter.com
Questions about General Product Safety Regulation:
productsafety@degruyterbrill.com

—

Olcay dedicates this book to Devin and Fusun, who have been steadfast pillars of support in his life, and to Chris, whose invaluable contributions made this book's publication possible.

Chris dedicates this book to Magil, for her encouragement and support, and to Olcay without whom this project would not have been started.

Preface

Welcome to the fascinating intersection of mathematics, biology, and ecology!

This book is intended primarily as a resource for teachers planning to teach their first introductory course on modeling in mathematical biology and/or ecology. This being said, it can also be used by students preparing to embark on an independent studies project in one of these fields; or, by researchers unfamiliar with the methods or software introduced who are seeking an accessible and quick introduction to one of the methods and/or software presented here; or, by curious biologists, ecologists, or mathematicians who may be unfamiliar with "the other side"; or, maybe, by the perpetual learner who is intrigued by the dynamics of living ecosystems. For each of the above, this book is designed to be an accessible introduction to the captivating landscape of biomathematics.

The approach used in this book takes advantage of technology in leading readers on a journey that bridges seemingly distinct fields through introductions to three methods and software platforms: compartmental models with Berkeley Madonna; agent-based models with NetLogo; and cluster analysis through self-organizing maps using an R Shiny app.

This is not intended to be a textbook (though it may be used as one), nor is it a purely mathematics book or one purely about deeper aspects of biology or ecology. It focuses on three selected ways in which the intersection of mathematics and biology (and mathematics and ecology) can be explored with the help of software. Moreover, the manner in which the content is presented makes it possible to use this book to help prepare for an introductory course at a wide range of levels, depending on the discipline within which the course is taught and the mathematical prerequisites for the course.

There are four chapters, the first of which presents the reader with a bit of background information followed by suggestions on how to get the most out of this book. The three core chapters introduce the three previously mentioned methods and software in a manner envisioned to be accessible to most.

https://doi.org/10.1515/9783111609560-203

Acknowledgments

A big thank you goes to Ute Skambraks, our content editor, for her encouragement, infectious enthusiasm and humor, and her knowledge in seeing this project through. We would also like to thank Nadja Schedensack for her efficiency and prompt help during Ute's sabbatical, and commend the de Gruyter design and production team for their excellent work in bringing this book to its final form.

https://doi.org/10.1515/9783111609560-204

Contents

Preface —— VII

Acknowledgments —— IX

1   **About the content** —— **1**
1.1   Introduction —— **1**
1.2   A bit of history —— **1**
1.3   A bit about the core chapters —— **2**
1.4   Ways in which this book can be used —— **2**
1.4.1   Undergraduate courses —— **4**
1.4.2   Beginning graduate courses —— **5**
1.4.3   Individualized research projects —— **6**
1.5   Where does this all lead to? —— **6**

2   **Compartmental modeling** —— **8**
2.1   Introduction —— **8**
2.2   Compartmental models in general —— **8**
2.3   Berkeley Madonna preliminaries —— **9**
2.4   An infectious-disease model —— **12**
2.4.1   Setting the stage —— **12**
2.4.2   Constructing the flowchart —— **13**
2.4.3   Solving the system —— **16**
2.4.4   Inserting sliders —— **16**
2.4.5   Some more graphics features —— **16**
2.4.6   Exploratory exercises —— **20**
2.4.7   Extensions and other ideas —— **22**
2.5   A predator–prey model —— **25**
2.5.1   Setting the stage —— **25**
2.5.2   Constructing the flowchart —— **26**
2.5.3   Solving the system —— **27**
2.5.4   Exploratory exercises —— **31**
2.5.5   Extensions and other ideas —— **33**
2.6   Selected resources —— **35**
2.6.1   Berkeley Madonna —— **35**
2.6.2   Background and theory —— **35**
2.6.3   Research and ideas —— **39**

3   **Agent-based modeling** —— **46**
3.1   Introduction —— **46**
3.2   Agent-based models in general —— **46**

3.3   The ODD protocol — 47
3.4   NetLogo preliminaries — 48
3.4.1   Key features — 49
3.4.2   Key components — 50
3.5   An infectious disease model — 53
3.5.1   Setting the stage — 53
3.5.2   Preparing the canvas and the code — 55
3.5.3   Monitors and plots — 62
3.5.4   Simulations and observations — 63
3.5.5   Exploratory exercises — 63
3.5.6   Extensions and other ideas — 69
3.6   Coding tips — 71
3.7   A predator–prey model — 73
3.7.1   Setting the stage — 73
3.7.2   Preparing the canvas and the code — 74
3.7.3   Simulations and observations — 82
3.7.4   Exploratory exercises — 82
3.7.5   Extensions and other ideas — 86
3.8   Selected resources — 88
3.8.1   NetLogo — 88
3.8.2   Background and theory — 89
3.8.3   Research and ideas — 94

4   Self-organizing maps — 106
4.1   Introduction — 106
4.2   Self-organizing maps and clustering — 106
4.2.1   The general idea — 106
4.2.2   The competitive learning algorithm — 107
4.3   RStudio — 108
4.4   The self-organizing map app — 109
4.4.1   Running the app — 110
4.4.2   App preliminaries — 110
4.5   Fisher's iris data — 115
4.5.1   Mapping plots — 117
4.5.2   Analysis plots — 120
4.5.3   Exploratory exercises — 123
4.5.4   Extensions and other ideas — 124
4.6   A bit more about this app — 125
4.7   A fish example — 126
4.8   Selected resources — 127
4.8.1   R, RStudio, and Shiny — 127
4.8.2   Open source data sets — 128

4.8.3   Background and theory —— 129
4.8.4   Research and ideas —— 133

Bibliography —— 139

Index —— 147

1 About the content

1.1 Introduction

This chapter provides a brief and select history of the fascinating field of biomathematics and those who developed the field; the structure of the following three core chapters; suggestions on how this book might be used; and a glimpse of where the ideas introduced might lead the interested and motivated learner.

1.2 A bit of history

For the interested, a very extensive resource on bibliographies of mathematicians can be found on the University of Saint Andrews MacTutor website, [82]. Detailed bibliographies and contributions of some of the individuals mentioned here (and many, many others) can be found on this website. Here is a selection of some notable discoveries and contributions to the applications of mathematics to biology and ecology.

A look into the past reveals that evidence of applications of mathematics to biology (and ecology) dates back to at least the 13th century when in 1202 Leonardo of Pisa (1175–1250), better known as Fibonacci, introduced the world of mathematics to his well-known sequence of numbers. This sequence is now called the Fibonacci sequence, and Fibonacci discovered it through a mathematical model he introduced for rabbit population growth. Over the years since, a wide range of examples of this sequence occurring in Nature have been found, and this sequence was also found to reveal the well-known and commonly occurring golden ratio (see, e. g., [22] or [98]).

The next notable event occurred in the 18th century courtesy of Daniel Bernoulli (1700–1782). It is worth mentioning that Bernoulli was best known for his work in hydrodynamics. However, in 1760, he wrote an article on what is considered the first mathematical model in epidemiology. The article describes smallpox's effect on the human population. This started the ball rolling in epidemiology (see, e. g., [11]). Also in the 18th century, Thomas Malthus (1776–1834) published his 1798 essay on human population growth; see [13] and [83], which began conversations on modeling population growth.

The 19th century saw further discoveries and the development of more population growth models. A notable contributor during this period was Pierre Verhulst (1804–1849) who formulated the logistic growth model, and published an article on this in 1845 (see, e. g., [32]). The ideas and works of Malthus and Verhulst on population growth were gradually built upon, and in the 20th century, Alfred Lotka (1880–1949) and Vito Volterra (1860–1940) independently discovered what are now referred to as the Lotka–Volterra predator–prey equations in 1920 and 1926, respectively. This model was extended and improved upon by others over time (see, e. g. [16] and [63]).

It was in the 1930s that Nicolas Rashevsky (1899–1979) developed a program in Mathematical Biophysics at the University of Chicago. He was also instrumental in establish-

https://doi.org/10.1515/9783111609560-001

ing The Bulletin of Mathematical Biophysics, now called The Bulletin of Mathematical Biology. Much of Rashevsky's work focused on modeling at the cellular level and he is credited by many as being the founder of mathematical biology as a recognized field in itself (see, e. g., [1] and [58]). It was not long after Rashevsky's contributions that there was an explosion in contributions to applications of mathematics to a wide range of areas in biology and ecology, accelerated no doubt by the advent of computers and specialized software.

The power of computers also contributed to the growing availability of large data and the ability to analyze these data more efficiently and effectively. Methods for identifying clusters in large data date back to as early as 1911 in applications to anthropology (see, e. g., [20] and [123]). Clustering (and the accompanying analysis) is now considered to be an important exploratory tool in looking for ideas prior to building models (where appropriate), hence its inclusion in this book.

Much more can be said, and asked, about mathematical biology (and ecology). For example, in [81] answers to the question "What has mathematics done for biology?" are explored. Then, in [102] answers to the question "Why is mathematical biology so hard?" are sought, and then some ideas on "Getting started in mathematical biology" are offered in [60].

1.3 A bit about the core chapters

These three chapters are structured in a manner that addresses four main themes for each of the three methods covered:
1. The method in question is introduced.
2. The software platform that is used to apply the method in question is introduced.
3. Two examples illustrating applications of the software platform are provided, along with suggested exploratory open-ended exercises and ideas for extensions.
4. A collection of selected resources on background theory and past research for ideas are provided for the interested reader. The bibliographies contained in these resources will serve as a source of ideas.

It should be mentioned here that while the Windows platform was used for all of the examples presented, the presentation remains valid for the Macintosh operating system.

So, how might this book be useful to a teacher of mathematics, or biology, or ecology who has decided to (or has been asked to) teach their first introductory course on biomathematical modeling?

1.4 Ways in which this book can be used

As will be seen in each of the last three articles cited previously, biomathematics along with all its foundational mathematics and underlying biological and/or ecological theory can be (and truly is) a very intense and complex field, which (eventually) requires fairly advanced foundations in these three areas. But it turns out that even if the hard mathematics is avoided at the introductory stage, with the help of software, its applications can still be quite informative (and possibly fascinating and fun) to the newcomer. This is the aim of this book—to provide ideas on how to introduce newcomers to the field of biomathematical modeling without overwhelming them with the mathematics (and/or statistics).

Unlike the traditional book on biomathematics, this book provides an accessible window into the field for the uninitiated without digging deeply into the mathematics, and to a large extent, the biology or ecology. The use of software makes this possible, and the goal of this book is to spark an interest in the field; in the curious teacher, student, or researcher. Also, this book is intentionally shorter than traditional works on the subject. Readers are unlikely to become overwhelmed, and will likely leave knowing enough to get started on applying the ideas and software covered to their classrooms or their own projects. It is also hoped that readers will end up with enough intriguing and/or exciting questions that will motivate further explorations into this important and remarkably diverse field.

As mentioned in the preface, this is not intended to be a textbook on the intricacies of the underlying mathematics (or biology, or ecology) behind biomathematical modeling. It does, however, contain ideas on presenting biomathematical modeling to those who are new to the field (and possibly with limited backgrounds in the mathematical sciences) in a nonintimidating yet informative and functionally effective manner.

So, this book is envisioned to be an instructor resource for a project-based and collaborative-learning course that introduces students to one or more of the three methods included. This means that aside from attending introductory lectures by the instructor, students would spend the majority of their remaining class time on their computers exploring possibilities through simulations while simultaneously engaging in vigorous and constructive student-to-student discussions (with periodic Socratic style interjections by the instructor).

Here are some possible settings for which a course based on ideas from this book could be designed and offered. In all scenarios outlined, it is important to remember that the finer details of course design should reflect both the nature/background of the target audience (the students) as well as the content and style of delivery the instructor is most comfortable with (or, at least, willing to experiment with). This being said, a collaborative and inquiry-based learning style class is strongly encouraged, with students being organized into groups of three to five students depending on class sizes and instructor workload considerations.

1.4.1 Undergraduate courses

While the mathematical prerequisites for the approach followed in this book is minimal, it is recommended that a fairly high bar on critical thinking and problem solving skills be imposed. In some cases, this may mean at least a first-semester calculus prerequisite, in others it may mean a class-level prerequisite (such as a minimum of a junior or senior undergraduate standing at U. S. institutions).

Assuming a class of suitably motivated students with adequate critical thinking and problem-solving skills, an instructor might choose one of the following:

– *Focus on depth* with respect to the modeling process through modules addressing a single area. For example, one possibility would an in depth development and analysis, including extensions, of infectious disease models using compartmental or agent-based modeling. The same could be done for population related models. Another possibility could be a sequence of multivariate data analysis exercises through clustering that target specific questions of interest.
– *Provide breadth* by covering one or two applications (depending on time availability) for each method introduced in the book so as to provide a functional foundation and a flavor of the three methods covered in the book.

The first of these lends itself to a rigorous one-semester introduction to the foundational ideas of modeling or data analysis, as well as a possible a sequence of three such one-semester courses (one for each method). The second lends itself to a one-semester "gentle introduction" to the methods and ideas covered in the book.

In keeping with the flexibility intended for this book, instructors can use the exploratory exercises and ideas for extensions provided at the end of each of the three core chapters to design modules that meet with their interests. At this level, it is recommended that modules involve a collection of fairly short and ideally open-ended guided inquiry-based learning exercises that can lead to lively discussions, and to interesting conclusions and/or further questions.

A three to four week module might take on the following form. Suppose, for example, the plan is to introduce students to constructing and analyzing an infectious disease compartmental model with Berkeley Madonna:

1. Begin with an introductory lecture on the underlying biology, epidemiology, and dynamics of a generic infectious disease (maybe provide students with a packet of brief notes for future reference).
2. Using the generic infectious disease example in Section 2.4, provide a classroom demonstration of constructing an SIR compartmental model using Berkeley Madonna, and have students follow along on their own computers.
3. Using one or more of the questions posed in Section 2.4.6, have students practice exploring the effects of varying parameter values and then make conjectures based on their observations—emphasize the open-endedness of these exercises.

4. Have a class discussion on conjectures arrived upon, and individual or group ratio-nales behind the proposed conjectures.
5. Have further class discussions on any differences (and/or similarities) between con-jectures arrived upon.
6. Have yet another class discussion on how, after conducting such analyses, the soundness of conjectures arrived upon might be informally assessed.
7. Choose a reasonably simple extension of the SIR model from the list provided in Sec-tion 2.4.7 and have students go through the process with minimal instructor input—refer all questions raised to the class as a whole. In the case of persistent difficulties, pose Socratic style questions rather than providing answers.
8. If so desired, close the module with a discussion on the mathematics behind the models constructed.

At this point, if the whole semester is to be devoted to constructing infectious disease compartmental models with Berkeley Madonna, have students choose a particular in-fectious disease and an appropriate model for a term project (see Section 2.4.7). Again, encourage discussions and have students practice posing meaningful questions as well as expose them to further Socratic style guiding questions.

1.4.2 Beginning graduate courses

Such a course could be offered to mathematics majors as an introduction to biomathe-matical modeling. It would also serve as a useful first course for advanced undergradu-ate or beginning graduate students from other disciplines, such as biology, wildlife ecol-ogy, etc.

The ideas expressed in the previous section for an introductory undergraduate course will work well in this case, too, but with higher expectations from students. One possibility for such classes would be to include an individually designed term project as one of the exercises. This can result in a very enjoyable class experience for both the students and the instructor. However, there will be times when courses designed in this manner can take off on an alarming tangent—the instructor will have to pay very close attention to the pulse of the class as a whole, as well as individual student progress through formative assessment methods.

For example, consider a course in which students are introduced to modeling wildlife population dynamics through agent-based models with NetLogo:
1. Start with a lecture on a generic two-species predator–prey model, avoiding any underlying mathematics but emphasizing relevant underlying theory from ecology. Preferably, introduce ideas through Socratic style questions.
2. Introduce students to the ODD protocol for agent-based modeling (see [51]), tying in discussions from 1.
3. Introduce students to preparing a flowchart that completely (or, at least satisfacto-rily) describes the dynamics of a predator–prey system.

4. Introduce students to the basics of coding, emphasizing the value of the NetLogo dictionary and using the example in Section 3.7 for a class demonstration. Have students follow along on their own computers.
5. Have students perform simulations and develop conjectures (see Section 3.7.4 for ideas).
6. Engage the class in discussions on conjectures arrived upon, about differences in opinions/strategies, and similarities.
7. Have students extend the model and then engage in more class discussions (see Section 3.7.5 for ideas).

At this point, students may be asked to pursue a term project of their choosing, preferably not too different in level of difficulty from the generic predator–prey model. Much would depend on the class level and student aptitudes.

Once again, the use of open-ended guided inquiry-based learning exercises that can lead to lively discussions, and to interesting conclusions and/or further questions is encouraged.

1.4.3 Individualized research projects

This would be for students or researchers who intend to embark on a particular research project and who may benefit from one or more of the methods and software introduced in this book. In essence, this person becomes both the teacher and the student.

Either one of the strategies outlined in Sections 1.4.1 and 1.4.2 would serve as a effective way to get started. Then ideas, extensions, and resources provided at the ends of each of Chapters 2–4 can be used to help decide on next steps.

1.5 Where does this all lead to?

It may be said that biology, at its core, is a dance of patterns, molecular interactions, ecological dynamics, and evolutionary processes—all governed by mathematical principles. From the spirals of a seashell to the branching of blood vessels, nature whispers its secrets in mathematical notation. Biomathematics provides a means of exploring these while searching for answers that play a crucial role in understanding and giving rise to methods useful for predicting various biological phenomena from a safe and ethical distance.

There is much more to biomathematics than can possibly be covered in a single book in any meaningful manner. Beyond what is covered in this book, here are just a few real-world areas of applications that can be explored with biomathematical modeling.

Epidemiology and disease modeling: Biomathematics helps model the spread of infectious diseases, predict the severity of outbreaks, and assess the impact of interventions (e. g., vaccination strategies, quarantine measures). In fact, it was instrumental during the COVID-19 pandemic (and past epidemics) in understanding transmission dynamics and optimizing public health responses.

Population dynamics: Understanding population growth, birth rates, and mortality rates is essential for conservation efforts, resource management, and sustainable development. Biomathematics includes modeling the dynamics of population systems that involve species ranging from endangered animals to invasive pests.

Ecology and ecosystems: Biomathematics helps analyze predator–prey interactions, food webs, nutrient cycling in ecosystems, and much more. It informs conservation strategies, such as protecting biodiversity and managing natural habitats.

Genetics and evolution: Mathematical models help scientists explore genetic inheritance, mutation rates, and evolutionary processes. They shed light on how traits spread through populations over generations.

Neuroscience and brain modeling: Biomathematics contributes to understanding neural networks, synaptic connections, and brain function. It aids in modeling brain diseases and optimizing treatments.

Pharmacokinetics and drug dosage: Biomathematics models drug distribution, metabolism, and elimination in the body. It guides drug dosage recommendations for patients.

Physiology and biomechanics: Mathematical models describe physiological processes, such as heart rate regulation, oxygen transport, and muscle contraction. Biomechanics studies movement, forces, and stresses in biological systems (e. g., analyzing gait patterns or bone fractures).

Environmental toxicology: Biomathematics assesses the impact of pollutants and toxins on ecosystems and human health. It predicts pollutant concentrations and their effects over time.

Bioinformatics and genomic analysis: Computational tools in available biomathematics are used to analyze DNA sequences, predict protein structures, and identify genetic variations. They provide methods useful in personalized medicine and disease diagnosis.

Agriculture and crop modeling: Biomathematics helps in optimizing crop planting schedules, predicts yields, and assessing the impact of climate change on agriculture. It helps design efficient irrigation systems and pest control strategies.

In summary, biomathematics bridges mathematics and biology (and ecology), allowing the unraveling of some of the mysteries of life and enabling informed decisions across diverse fields. The resources listed at the end of each of the three core chapters are intended to provide useful starting points for those interested in pursuing further studies in the areas introduced, as well some not covered.

2 Compartmental modeling

2.1 Introduction

This chapter expands on the presentation in [4], while retaining a focus on demonstrating the ease with which the differential equations solver *Berkeley Madonna* [84] can be used to create and solve first-order *compartmental models*. Berkeley Madonna was created at the University of California, Berkeley by Robert Macey, George Oster, and Timothy Zahnley and formally incorporated in 1998. As described on its website [15],

> Berkeley Madonna is an incredibly fast, general-purpose differential equations solver. Its graphical interface provides an intuitive platform for constructing complex mathematical models with ease using symbols rather than writing equations. The software provides a suite of graphical tools for plotting your results.

In a nutshell, the construction and solution of a compartmental model in Berkeley Madonna is accomplished by constructing a flowchart that defines the manner in which model variables are related and/or interact. This chapter focuses on an introductory exploration of compartmental models using Berkeley Madonna.

2.2 Compartmental models in general

As will be evident from the examples described in Sections 2.4 and 2.5, it is not necessary to have a formal background in differential equations to begin using Berkeley Madonna. However, it is worth providing a brief mathematical description of what a compartmental model looks like in general.

Let $X_1, X_2, \ldots, X_n$ be continuous time-dependent variables and let $F_1, F_2, \ldots, F_n$ each denote functions of one or more of $X_1, X_2, \ldots, X_n$. Also, allow for the possibility that each of these functions may include one or more parameters. Then, in *certain* scenarios, a system of first-order differential equations of the form shown in the System (2.1) is called a *compartmental model*:

$$
\begin{aligned}
\frac{dX_1}{dt} &= F_1(X_1, X_2, \ldots, X_n), \\
\frac{dX_2}{dt} &= F_2(X_1, X_2, \ldots, X_n), \\
&\ \vdots \\
\frac{dX_n}{dt} &= F_n(X_1, X_2, \ldots, X_n).
\end{aligned}
\tag{2.1}
$$

For the examples in Sections 2.4 and 2.5, the variables $X_1, X_2, \ldots, X_n$ represent the sizes of different populations, or sizes of subsections of a given population determined by

https://doi.org/10.1515/9783111609560-002

some characteristic of interest at a given time. However, in general, these variables can represent a wide range of other quantities.

Each variable can then be thought of as representing a quantity associated with a *compartment* in the system under study. The derivatives on the left side of System (2.1) represent the rates of change with respect to time of the quantities within each compartment. The way in which the functions $F_1, F_2, \ldots, F_n$ in a compartmental model are defined is important, as these describe the mechanisms that drive changes in the system under study. For applications of interest, these functions are typically *nonlinear* and the equations in System (2.1) are always *coupled*—at least one of the equations contains more than one of the variables $X_1, X_2, \ldots, X_n$. It is in constructing the functions $F_1, F_2, \ldots, F_n$ that Berkeley Madonna shines for the novice modeler.

2.3 Berkeley Madonna preliminaries

The Berkeley Madonna website provides all needed instructions for downloading and installing the Berkeley Madonna platform. When Berkeley Madonna is started up, a blank gray screen titled Berkeley Madonna opens up with a menu-bar at the top. Of preliminary interest at this point is the File menu option New Flowchart Document shown in Figure 2.1. The other two items in the menu-bar that will be of interest are Graph and Parameters; see Figures 2.2 and 2.3.

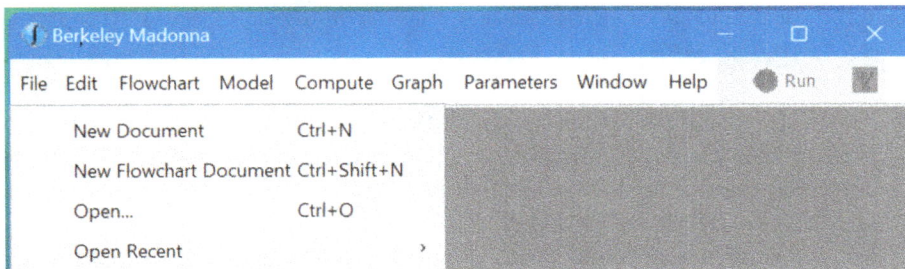

Figure 2.1: Creating a new flowchart from the File menu. The remaining menu options are self-explanatory.

To get started, click on the New Flowchart Document option under File in the main menu-bar; see Figure 2.1. This opens a blank flowchart; see Figure 2.4. The flowchart window in the center of Figure 2.4 is where flowcharts are constructed and the panels on either side of this are where the corresponding model details needed for the numerical solution appear as the flowchart for the model is constructed.

Before continuing, hover the cursor over each of the yellow icons in the menu-bar at the top of the flowchart window to see what each icon is named; see Figure 2.4. Four of these are of initial interest.

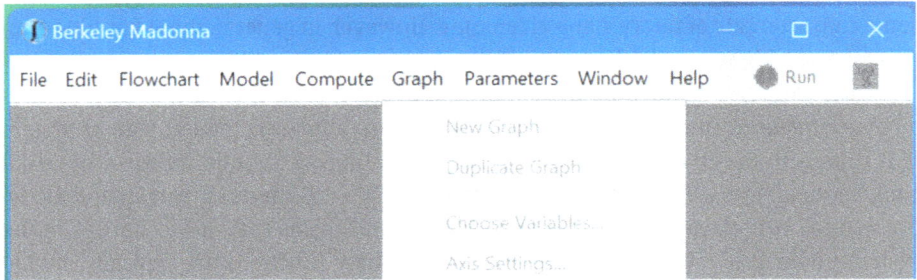

Figure 2.2: Options within the Graph menu-item that will be of interest are New Graph, Choose Variables, and Axis Settings.

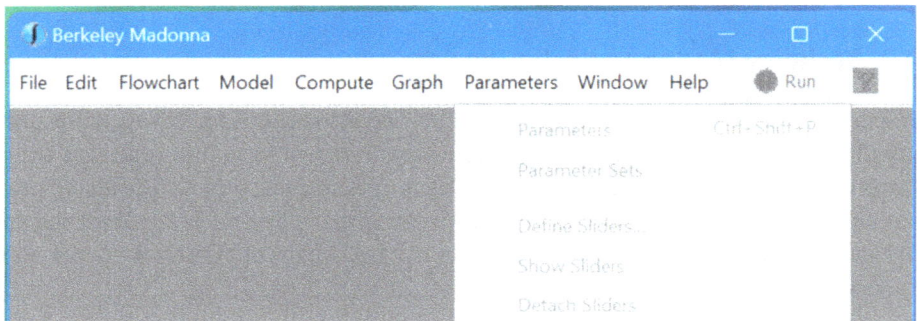

Figure 2.3: Options within the Parameters menu-item of interest are Parameters, Define Sliders, Show Sliders, and Detach Sliders.

- reservoir: These are used to represent distinct populations, or distinct subpopulations within a larger population. In the language of compartmental models, these are the *compartments* within a compartmental model and represent the variables for which the model is to be solved.
- arc: These are used to identify the reservoirs that contribute to flows into and out of compartments.
- flow: These are used to define the manner in which the size of each compartment changes with respect to time. An outflow from a reservoir indicates a decrease in size and an inflow indicates an increase in size.
- global: This is used to define global parameters, or any other global constant quantities needed in the model construction and subsequent solution process. Once one or more global quantities are defined, a new button labeled Globals appears next to the Flowchart button above the flowchart. This can then be used to make changes to the defined set of global quantities.

Illustrations of how to use these four tools in the flowchart window are given in the construction of the model in the following example.

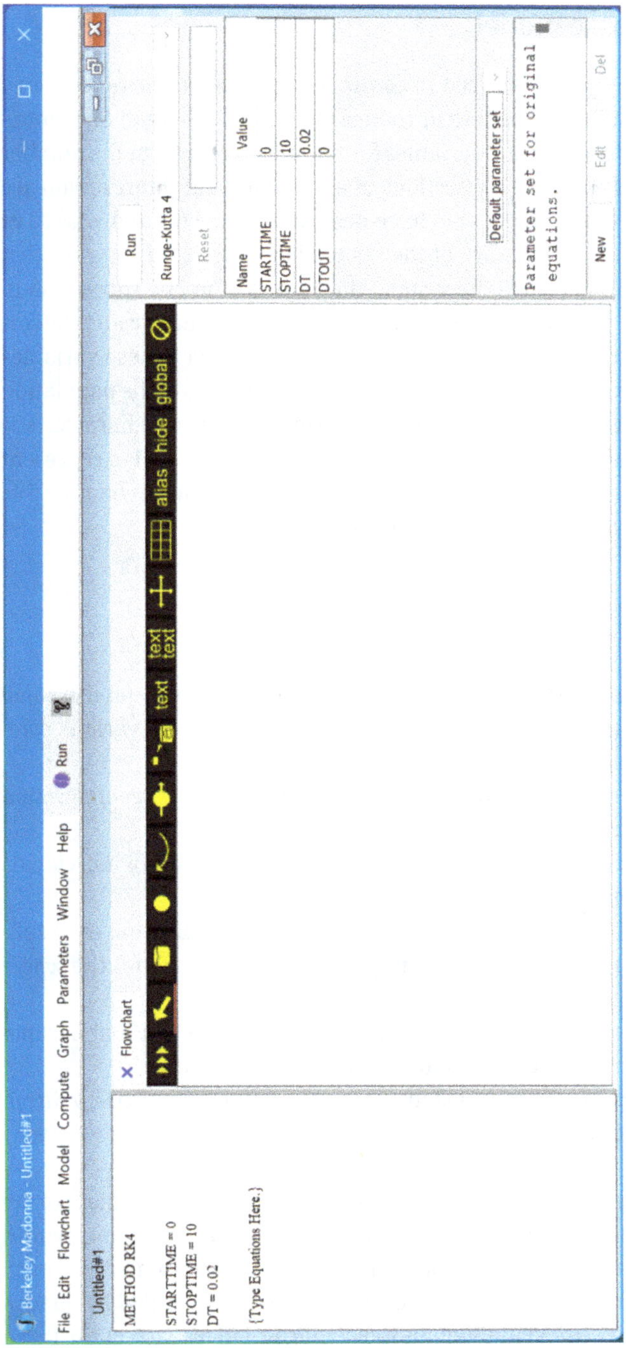

Figure 2.4: A blank flowchart displaying the model construction buttons of interest and more.

2.4 An infectious-disease model

The model used for this example is referred to as the *Susceptible–Infected–Recovered* (SIR) model. This model is commonly used to introduce modeling the spread of communicable diseases. At time $t \geq 0$ in days, the variables for this model represent the number (or proportion) of individuals in those subsections of a population of interest who are susceptible to the disease in question (S), who have become infected (I), and who have recovered from the disease (R). The nature of the system being modeled requires that members of the population never leave the system; they move from one subsection to the next. So S, I, and R describe the *state* of each subsection of the population and the corresponding *state variables* describe a system in *equilibrium*. Moreover, these variables all have the same dimensions, representing the sizes of the subsections of the population of interest at any time t with each subsection being determined by some characteristic of interest. For purposes of this example, the variables S, I, and R are scaled to represent proportions of the whole population rather than counts within each compartment—this makes them dimensionless (nondimensional) variables.

2.4.1 Setting the stage

The next phase in constructing the model for the system in question is to establish some assumptions and further notation. These will aid in putting together the various components that describe the mechanics of the system being modeled.

– For the time period in question there are no births or deaths, and neither emigration nor immigration occur, so $S + I + R = 1$.
– Susceptible individuals become infected and then recover, so the flow of individuals occurs from S to I and then from I to R.
– Instantaneous mixing of the population occurs. That is, individuals from each subsection of the population are *always* uniformly distributed throughout the *whole* population.
– Susceptible individuals who come in contact with an infected individual become infected at some constant rate per unit time, the *transmission rate α*.
– Infected individuals recover at some constant rate per unit time, the *recovery rate β*.
– Recovered individuals gain immunity from the disease and do not get reinfected.

With the help of these assumptions, the flow of individuals from one compartment to the next can now be defined. The variables S, I, and R represent the compartments within the system being modeled, and in Berkeley Madonna, these are represented by *reservoirs*. The underlying mechanics for this particular model can be described by quantifying the rate (per unit time) at which individuals move into, or out of, each of the three compartments. In Berkeley Madonna, *flows* combined with *arcs* and *globals* provide a means of defining the direction and rate at which members from each compartment move—arcs are used to identify which variables contribute to this movement, and globally defined parameters provide the relevant rates of change per unit time.

Starting with susceptible individuals, because of instantaneous mixing it can be argued that the product SI provides a measure of the proportion of interactions between the susceptible and infected individuals that occur at any given time. Then, subject to the infection rate a, at any given time the proportion of susceptible individuals will decrease through infections by the quantity aSI per unit time.

Moving to individuals who have recovered from the disease subject to the recovery rate β, at any given time the rate at which the proportion of recovered individuals increases per unit time will be βI.

Finally, the change in the proportion of infected individuals per unit time is determined by the inflow of newly infected individuals and the outflow of recovered individuals. Thus, the rate at which infected individuals increase (or decrease) per unit time is given by $a\,SI - \beta I$.

2.4.2 Constructing the flowchart

Recall that a `reservoir` represents a compartment and a `flow` provides the direction in which members from each compartment move. Begin the flowchart construction by inserting these in the flowchart window as shown and described in Figures 2.5 and 2.6.

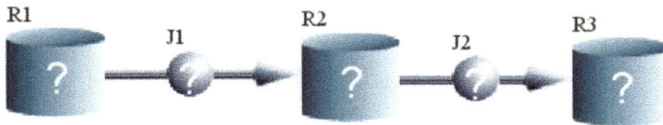

Figure 2.5: First insert the reservoirs by dragging three of them onto the flowchart window. Next, insert flows by first clicking on the `flow` icon and then drag flows from the first to the second reservoir, and then from the second to the third reservoir.

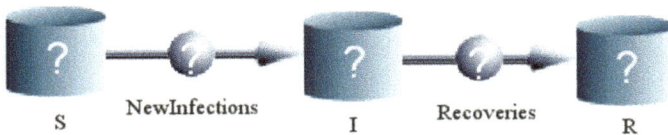

Figure 2.6: Rename the reservoirs and flows one by one by double-clicking on the existing names and then entering the new names in the text dialog windows that pop up. Once this is done, move the names by dragging them to desired locations.

To set initial values for each reservoir, double-click on the "?" symbol on each reservoir and enter the initial values for each; see Figures 2.6 and 2.7. For this example, use $S_0 = 0.9999$, $I_0 = 0.0001$, and $R_0 = 0$.

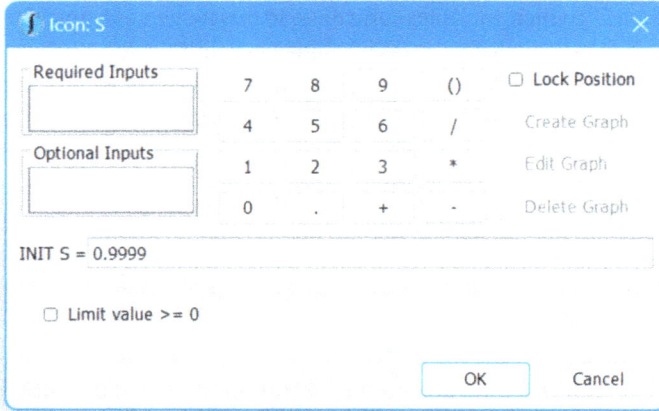

Figure 2.7: Setting initial values. Double-click on the S reservoir to open the S-icon window, then enter the initial proportion for the susceptible population. Repeat the process for the I and R reservoirs to get the flowchart shown in Figure 2.8.

Figure 2.8: The flowchart appearance once the initial proportions for S, I, and R have been entered. Notice that the "?" symbols on the reservoirs have gone away.

The first step in defining the nature of flows from one reservoir to the next is to provide the values for the parameters α and β. To do this, click on the global button on the flowchart menu-bar (see Figure 2.4), and then enter the values as shown in Figure 2.9. Next, go back to the flowchart window and insert the relevant arcs from the reservoirs to the respective flows as shown and described in Figure 2.10.

Figure 2.9: Enter parameter values α and β as global quantities. For this example, suppose that for each interaction a new infection occurs every two days, meaning $\alpha = 0.5$. Also, suppose it takes approximately 3 days for an infected individual to recover; use $\beta = 0.33$.

Finally, double-click on each of the flow buttons and enter the appropriate computational formulas decided upon in Section 2.4.1; see Figures 2.11 and 2.12. Once all of this has been done, save the completed flowchart for this SIR model. Figure 2.13 provides a

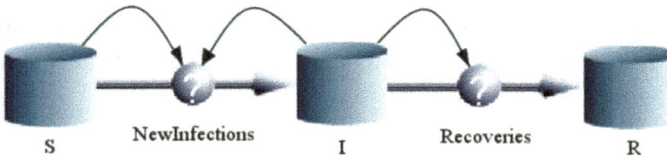

Figure 2.10: To insert an arc, click on the arc icon in the flowchart menu-bar and then drag an arc from one reservoir to a flow button. Since the proportion of new infections depends on both S and I, insert arcs from the S and I reservoirs to the flow out of the S reservoir. Since the proportion of recoveries depends on I, insert an arc from the I reservoir to the flow out of the I reservoir.

Figure 2.11: To define the flow out of the S reservoir, double-click on the NewInfections "?" symbol to open the NewInfections icon window. Notice that both S and I are required inputs and α and β are optional inputs. The proportion of new infections is αSI, so enter NewInfections = alpha*S*I.

Figure 2.12: To define the flow out of the I reservoir, double-click on the Recoveries "?" symbol to open the Recoveries icon window. The proportion of recoveries is βI, so enter Recoveries = beta*I.

complete picture of what appears on this flowchart and in the panels to the left and the right of it. The task now is to solve the system of equations in the left-hand panel that define the model.

2.4.3 Solving the system

The right-hand panel in Figure 2.13 provides relevant settings for the numeric solution of the system of equations provided in the left-hand panel. Before proceeding to the solution of this system, click on STOPTIME in the panel on the right and set this to 150 (*do not* click on the Reset button), and similarly set the time-step DT to 0.01—the smaller this is the smoother the solution curves will be.

Right below the Run button, notice that the default numerical scheme in Berkeley Madonna is the *fourth-order Runge–Kutta method*—leave this as is. Then click on the Run button to solve the system and obtain the solution curves; see Figure 2.14.

2.4.4 Inserting sliders

To include an additional useful feature on this graph, go to the main menu-bar, click on the Parameters button, and then select the Define Sliders option; see Figure 2.3. Then, in the Define Sliders pop-up window (see Figure 2.15), click on and add the parameter alpha, setting the minimum to 0 and the maximum to 1. Similarly, repeat the process to add a slider for the beta parameter and set its range from 0 to 1. The result is displayed in Figure 2.16.

Note that the range (Minimum and Maximum) for sliders does not need to be [0,1]. Choose a range that best fits the needs of the scenario under study. Also, once sliders are defined and inserted in a graph, they can be hidden (or detached) by selecting Hide Sliders (or Detach Sliders) in the Parameters main menu. Or they can be reinserted by selecting Show Sliders (or Reattach Sliders) in the Parameters main menu.

It is now possible to explore the dynamics of the epidemic for different parameter values by moving the sliders back and forth. To do this, drag the sliders or click on the blue arrows to the left of the sliders—solution curves are updated instantaneously. It may be that STOPTIME will have to be adjusted with changes to the parameter values.

2.4.5 Some more graphics features

The Readout icon in the menu-bar above the graphics window (see Figure 2.16) is useful for finding the coordinates of a point on a solution curve. Click on the Readout icon and move the cross-hairs to, and click on the desired point on the curve of interest. The coordinates will appear in the lower-left corner of the graphics window.

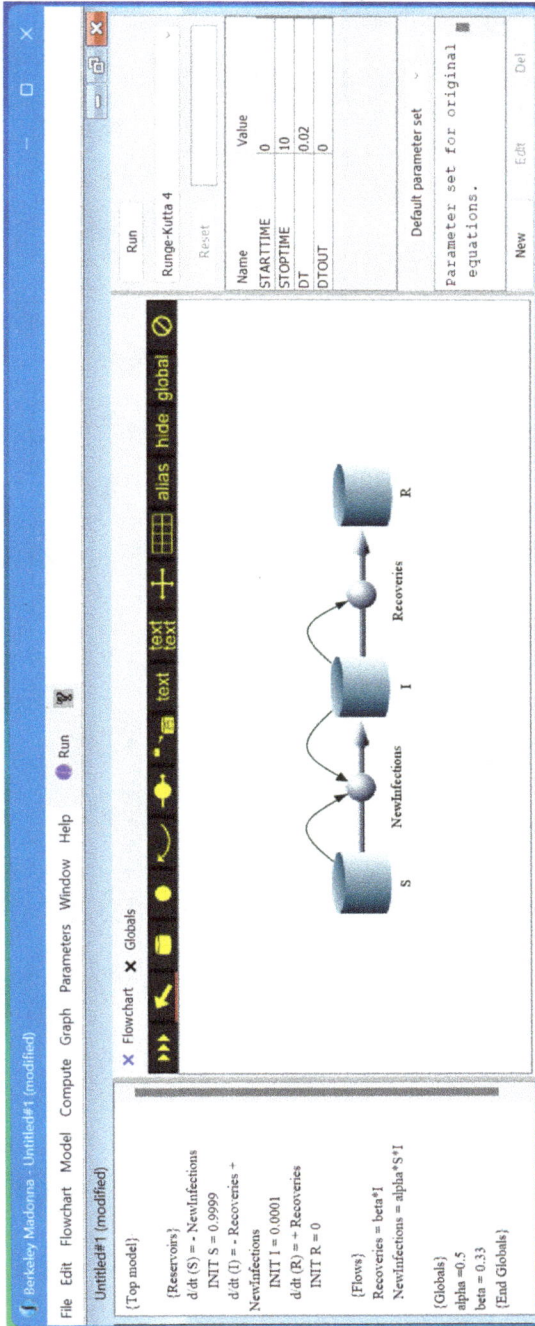

Figure 2.13: The completed flowchart for, and definition of the SIR model. The left-hand panel displays the model along with parameter settings and the right-hand side shows default values for time.

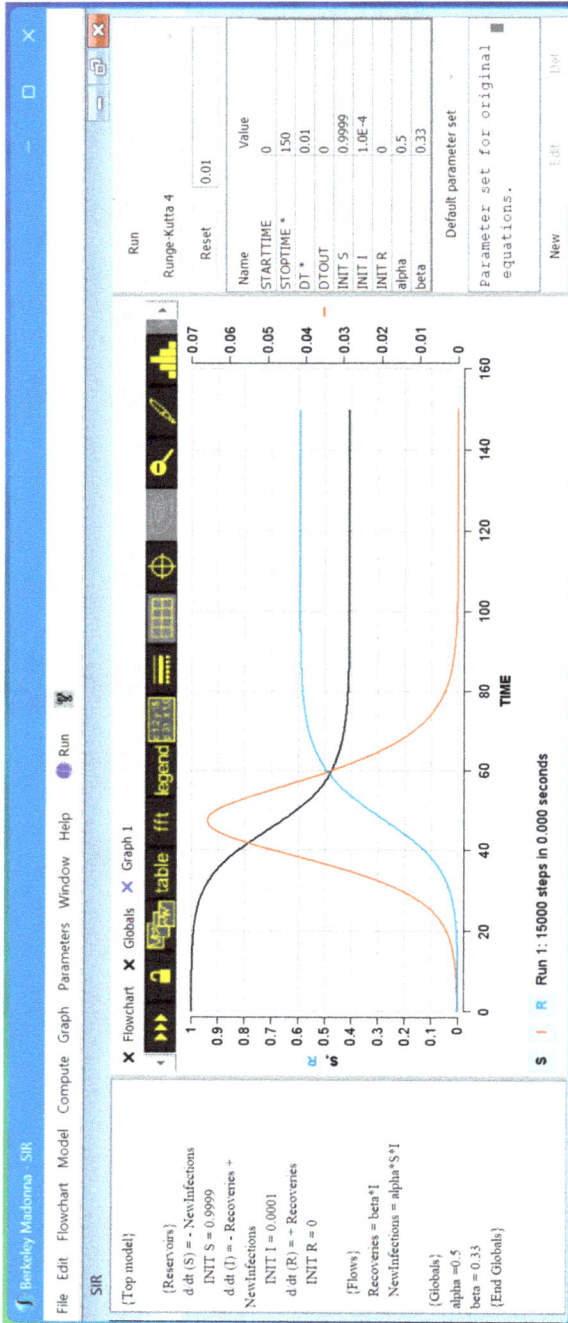

Figure 2.14: Solution curves for the SIR model. Change STOPTIME to 150 and DT to 0.01, but leave the numerical solution method as Runge-Kutta 4.

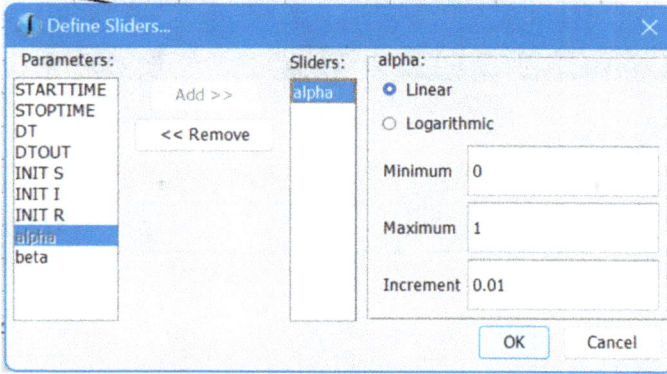

Figure 2.15: Defining and inserting a slider for the parameter alpha. A slider for beta is similarly inserted.

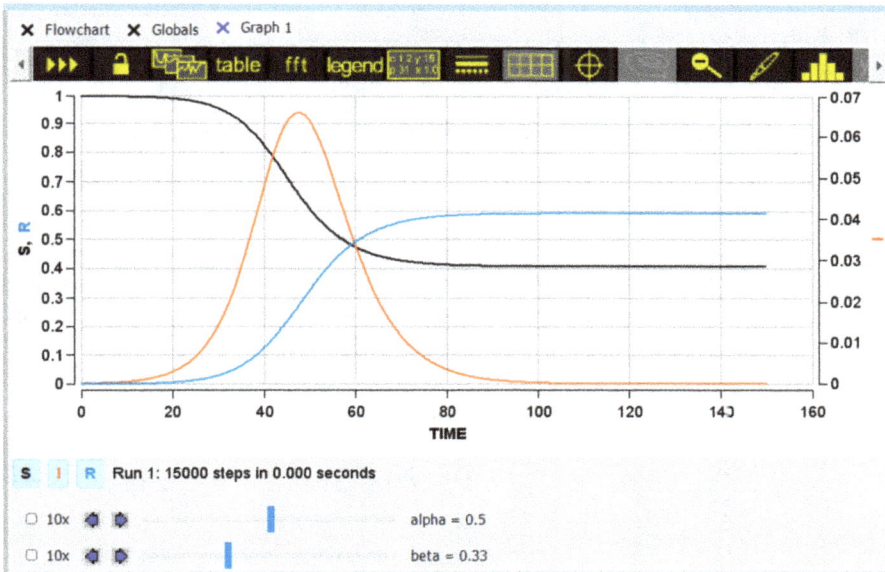

Figure 2.16: Graphs of the SIR solution curves with sliders for alpha and beta included.

Observe that in Figure 2.16 the vertical scale for the infected curve (I), indicated on the right-hand vertical axis is different from the vertical scale for susceptible and recovered curves (S and R, respectively). Axis settings can be changed, if so desired, by clicking on the Graph button in the main menu-bar and selecting Axis Settings to change the default settings. Figure 2.17 shows how the appearance of Figure 2.16 changes if the left and right vertical axes have the same scale. However, it is worth mentioning that user-defined axes settings (rather than leaving them in auto mode) will sometimes produce graphs that appear empty or incomplete.

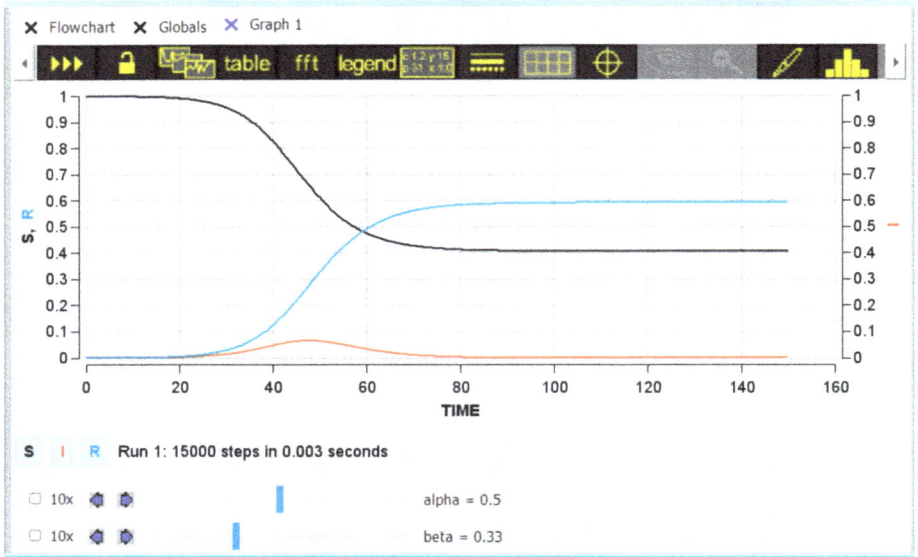

Figure 2.17: Graphs of the solution curves with sliders and the right-hand axes scale redefined.

2.4.6 Exploratory exercises

On careful examination of Figure 2.14 it will be noticed that the panel on the left provides details of the actual mathematical model itself, that is, a specific case of the generic SIR model:

$$\frac{dS}{dt} = -\alpha SI,$$
$$\frac{dI}{dt} = \alpha SI - \beta I, \tag{2.2}$$
$$\frac{dR}{dt} = \beta I,$$

with initial values being $S(0) = S_0$, $I(0) = I_0$, and $R(0) = R_0$. Observe that a direct implication of System (2.2) is

$$\frac{dS}{dt} + \frac{dI}{dt} + \frac{dR}{dt} = 0,$$

which is also a direct consequence of the first assumption of the system being modelled. Notice also that the proportions S, I, and R are dimensionless, so the parameters α and β both have units t^{-1}.

Thus far, the example presented here is along the lines of the discussion and model presented in [110]. Moving forward, having constructed the model and set the stage for solutions it now possible to easily and quickly perform simulations for different values of the parameters with the help of the sliders. Before continuing, reset the right-hand axis scale for the solution curves to auto as displayed in Figure 2.16.

Here are some exploratory exercises that suggest ways in which Berkeley Madonna might be used to perform an analysis of an SIR compartmental model for a particular infectious disease through simulations:

1. With the model being as defined, use the Readout tool to determine after approximately how many days will the number of susceptible and recovered individuals be the same. Be sure to deactivate the Readout tool once this is done—click on the icon again.

2. Suppose the recovery rate of infected individuals remains at β = 0.33. Since the parameter α determines the rate at which the disease spreads the α-slider can be used to see how the nature of the curves change when the value of α is raised or lowered. What (if any) might be a reason for changes in the value of α in a real-world setting?

3. With β = 0.33, for approximately what value of α will the proportion of susceptible and recovered individuals end up being approximately the same in the long run?

4. Now suppose the infection rate remains at α = 0.5. Since the parameter β determines the rate at which infected individuals recover, the β-slider can be used to see how the nature of the curves change when the value of β is raised or lowered. What (if any) might be a reason for changes in the value of β in a real-world setting?

5. With α = 0.5, for approximately what value of β will the proportion of susceptible and recovered individuals end up being the approximately same in the long run?

6. Play around with the values of *both* α and β and look for that pair of values for which the solution curve of I has a *nonzero* horizontal asymptote. Can you find a pair (α, β) that will produce such a curve? If yes, is this pair unique? If no, why not? If you cannot find such a pair, what might be a reason for this?

Notice that the scenario described in the last exercise above corresponds to that in which dI/dt in the System (2.2) approaches zero as $t \to \infty$. Suppose a particular infectious disease is introduced to a given population and that:

– Everyone in the population is free from infection and has never been exposed to the disease before;
– Nobody in the population has been vaccinated against the disease; and
– The spread of the disease cannot be controlled.

Consider the scenario in which the disease starts with $(S, I, R) = (S_0, I_0, 0)$ and $dI/dt = 0$. Then, at $t = 0$,

$$aS_0I_0 - \beta I_0 = 0 \quad \rightarrow \quad \frac{aS_0}{\beta} = 1.$$

Now define

$$\mathcal{R}_0 = \frac{aS_0}{\beta}. \tag{2.3}$$

For an infectious disease in such a scenario, $\mathcal{R}_0$ is referred to as the *basic reproduction number* (or *ratio*) for the disease and this provides an estimate of the number of susceptible individuals who will contract the disease from one infected individual. For example, if $\mathcal{R}_0 = 3$ then an infected individual will, on average, transmit the disease to 3 other individuals. If no one in the community has been vaccinated against the disease or is already immune to it, this process continues to repeat itself for each newly infected individual.

Depending on the $\mathcal{R}_0$ value, there are three possibilities that exist for the potential spread and/or decline of an infectious disease:

- If $\mathcal{R}_0 < 1$, then $dI/dt < 0$ and each existing infection causes less than one new infection. In this case, the spread of the disease will decline and eventually die out.
- If $\mathcal{R}_0 = 1$, then $dI/dt = 0$ and each existing infection causes one new infection. The disease will stay alive and stable, but there will not be an outbreak or an epidemic— the disease becomes *endemic*.
- If $\mathcal{R}_0 > 1$, then $dI/dt > 0$ and each existing infection causes more than one new infection. The disease will be increasingly transmitted between people resulting in a possible outbreak or *epidemic* before eventually declining.

Clearly, estimating the $\mathcal{R}_0$ value for an infectious disease is important, and finding conditions under which the value of this number is reduced (preferably below 1) would be very helpful.

7. Think about $\mathcal{R}_0$ and the current SIR model. Does it make sense to talk about an endemic disease when using the SIR structure to model the disease? Why or why not?
8. What might be some real-world strategies to ensure that $\mathcal{R}_0 < 1$ for the SIR model?

2.4.7 Extensions and other ideas

Based on the assumption that no births, deaths, emigration, or immigration occur, it is evident that this model is best suited for a closed community over a relatively short time period. This, combined with the assumption that reinfections do not occur due to acquired immunity, implies that the SIR model always predicts the eventual decline of a disease it is used to model. This may not be realistic, and for this reason other models might be preferred.

It turns out that the strategy just used in constructing the SIR model in Berkeley Madonna can be extended to other compartmental models for infectious diseases. Alternative models can be obtained by extending the SIR model through asking, and answering questions. For example:
–   What might be an alternative compartmentalization of the population?
–   How many compartments will there be in the proposed alternative (or extended) model? Will there be fewer or more?
–   What are expected roles that these compartments would play?
–   What will the direction of flow between compartments look like?
–   Will the number of parameters involved change from the number in the SIR model? If yes, will there be fewer or more? What will these be?
–   How will the compartments and parameters contribute to the definitions of flows out of and/or into compartments (think arcs)?
–   What might be useful/informative solution curve plots?

Once a model is constructed (in Berkeley Madonna), other questions can be asked. For example:
–   What does the system of differential equations look like? How does it differ from that of the SIR model?
–   For which parameters should sliders be included? Why?
–   What information does varying each parameter provide?
–   What is a meaningful goal of performing an analysis of the model through parameter variations?

The following selection of models can be constructed in Berkeley Madonna by extending the SIR model.
–   **SIRS** (Susceptible-Infected-Recovered-Susceptible): This is similar to the SIR model, except in that a recovered individual is assumed to gain short-term immunity and eventually moves back into the susceptible pool.
–   **SIRV** (Susceptible-Infected-Recovered-Vaccinated): This too is similar to the SIR model in all respects except in that members from the susceptible pool who gain at least short-term immunity through vaccination are accounted for. Individuals who have recovered are assumed to have gained at least short-term immunity.
–   **SIRD** (Susceptible-Infected-Recovered-Deceased): This is similar to the SIR model in all respects except in that infected individuals may recover and gain immunity, or they may become deceased.
–   **SIS** (Susceptible-Infected-Susceptible): This model assumes the absence of immunity (even after recovering from an infection), and that an infected individual who recovers moves back into the susceptible pool.
–   **SEIS** (Susceptible-Exposed-Infected-Susceptible): This is similar to the SIS model, except in that there is an in-between period of time when a susceptible individual

has been exposed to the disease and does not clearly display visible signs of infection. However, during this period of time, even though exposed individuals do not display visible signs of infection, they are assumed infectious.

- **SIRVD** (Susceptible-Infected-Recovered-Vaccinated-Deceased): This model combines elements of the SIR, SIRV, and SIRD models.
- **MSIR** (Maternal-Susceptible-Infected-Recovered): Here, the term "maternal" identifies individuals (typically babies) who have gained passive immunity from maternal antibodies. So, this model accounts for another form of immunity.
- **SEIR** (Susceptible-Exposed-Infected-Recovered): This too is similar to the SIR model, except in that there is an in-between period of time when a susceptible individual has been exposed to the disease but does not clearly display visible signs of infection. During this period of time, even though the exposed individuals do not display visible signs of infection, they are assumed infectious.
- **MSEIR** (Maternal-Susceptible-Exposed-Infected-Recovered): This is similar to the SEIR model, except in that there is a proportion of the population that has acquired immunity through maternal antibodies.
- **MSEIRS** (Maternal-Susceptible-Exposed-Infected-Susceptible): This follows the MSEIR model, except in that recovered individuals become susceptible again after a period of time.

It should be mentioned that this is just a partial list of possible compartmental models for infectious diseases. One could ask further questions. For example, to list just a few:

- Can births and deaths due to natural causes (those unrelated to infection) be brought into the picture?
- Would it be meaningful to bring in births and deaths due to natural causes into the picture?
- Does age play a role in infection rates? Would this be worth exploring?
- Does age play a role in recovery rates? Would this be worth exploring?
- Does age play a role in deaths due to infection? Would this be worth exploring?
- How might quarantining be brought into the picture? Would this be worth exploring?
- Can immigration and emigration be brought into the picture? Would this be worth exploring?

Of course, one must not forget that a big part of getting started on constructing any model is acquiring a good understanding of the underlying biology, or in this case, epidemiological background information. See Section 2.6.2 for a selection of resources on this.

2.5 A predator–prey model

For this example, consider modeling the population sizes of a single prey species and its sole predator species in a closed ecosystem. Let the continuous time-dependent variables X and Y represent the sizes of these two populations, respectively, where time $t \geq 0$ is measured in years. Unlike for the previous SIR model, the variables of a *predator–prey model* do not necessarily describe a system in equilibrium. However, each species does represent a compartment in the system.

2.5.1 Setting the stage

Before beginning the model construction process, first establish some assumptions.
- The two populations experience instantaneous mixing, meaning the two species are always uniformly distributed amongst each other.
- There is an unlimited source of food for the prey.
- In the absence of predation, increases in the prey population size per unit time occur through births, which are assumed to outpace natural mortality.
- Decreases in the prey population size occur solely through predation.
- Increases in the predator population size per unit time occur through births, which depend on the availability of prey.
- Decreases in the predator population size occur through natural mortality (which includes deaths through starvation).

The components that describe the dynamics of the system to be modeled can now be built.

First, consider changes in the prey population per unit time. Let b denote the birth rate per unit time for the prey. Then the prey population size will increase by the quantity bX per unit time. Next, denote the rate at which each predator–prey interaction results in a prey death by δ. Then the decrease in the prey population size per unit time as a result of predation will be δXY.

Moving to changes in the predator population size, let d denote the death rate for the predator per unit time. Then the decrease in the predator population size per unit time will be dY. Finally, recognizing that predator births depend on predator–prey interactions, denote the rate of growth in the predator population size due to the effect of each interaction with a prey by γ. Then the predator population size increase per unit time will be γXY.

This information can then be used to construct a flowchart that describes the predator–prey system in question, the process being similar to what was done for the previous SIR example.

2.5.2 Constructing the flowchart

There is one important point to note for this system. Unlike in the previous example, individuals from the two populations (compartments), X and Y, do not move from one to the other. This means that while *flows* into and out of each of the two reservoirs exist, flows between the two reservoirs are not drawn; see Figure 2.18. However, interactions between the two do occur and *arcs* identifying variables used in defining a flow may come from one or both reservoirs; see Figure 2.19.

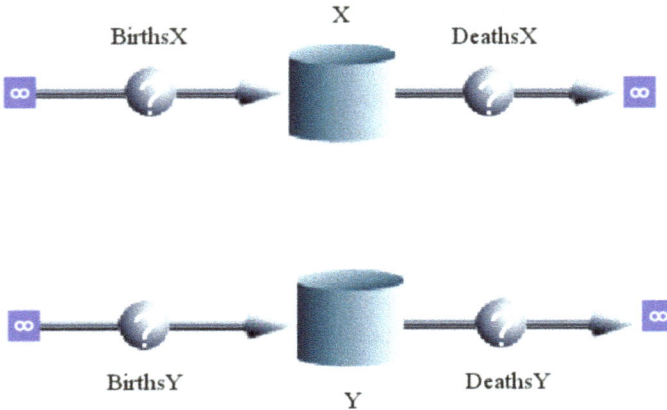

Figure 2.18: Reservoirs and flows for the predator–prey system. Initial population sizes are set at $X_0 = 25$ and $Y_0 = 15$.

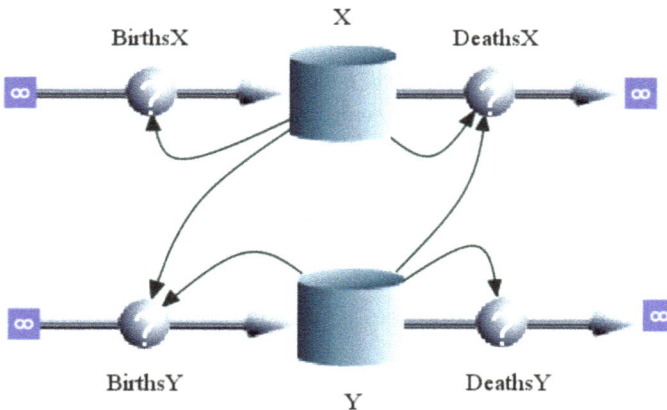

Figure 2.19: Arcs for the predator–prey system.

For purposes of this example, consider using the parameter values $b = 1, d = 0.5, \delta = 0.01$, and $\gamma = 0.0005$ as Globals, and consider initial populations of $X_0 = 25$ and $Y_0 = 15$ (refer back to Figures 2.7 and 2.9). Next, following the process described for completing the flowchart in the previous example, flows into and out of the reservoirs X and Y are defined (refer back to Figures 2.10–2.12). Then the completed flowchart describing the predator–prey system in question takes on the appearance of Figure 2.20.

2.5.3 Solving the system

Using a STOPTIME of 30, click on the Run button to get the solution curves; see Figure 2.21. Notice that the left-hand vertical axis of the graph represents the prey population size, the right-hand vertical axis represents the predator population size, and the scales of these two axes differ. If so desired, the two scales can be made the same as was done to get Figure 2.17 for the previous example. That is, in the Axis Settings pop-up window from the Graph main-menu, deselect the auto scaling boxes for the left and right vertical axes, and make the two scales the same.

Another common plot used for the analysis of such systems is the plot of the predator population size against the prey population size (Y against X). To get this graph, select New Graph from Graph in the main-menu; see Figure 2.2. Then, with the Graph 2 window activated, select Choose Variables from the Graph main-menu. Next, in the Choose Variables pop-up window, place X in the X-Axis and Y in the Y-Axis; see Figure 2.22. Click on OK and then, if so desired, open the Axis Settings window and customize the axes labels as shown in Figure 2.23. Finally, with the Graph 2 window activated, click on Run again. The resulting graph is shown in Figure 2.24.

Here is another nice feature. In the Graph 2 window (see Figure 2.24), hover the cursor over the yellow icons in the menu-bar above the graph to see the names of the button options. There are two buttons of particular interest:

- Overlay Plots: This enable multiple plots on the same axes.
- Initial Conditions: When this button is clicked on and the cursor is moved over the graphics window the cursor takes on the form of cross-hairs as for the Readout button. As the cursor is moved around its position in the coordinate axes is given in the lower-left side of the graph. Clicking on the mouse at a particular point marks that point as a new pair of initial values for X and Y and the corresponding solution curve appears.

The previously mentioned Readout button remains useful for estimating points on solution curves.

Click on both the Overlay Plots and the Initial Conditions icons and move the cursor onto the graphics window. Pick a location on the coordinate axes and click the mouse. The new overlaid graph (see Figure 2.25) represents the graph of the system with initial conditions corresponding to the point on the coordinate axes where the

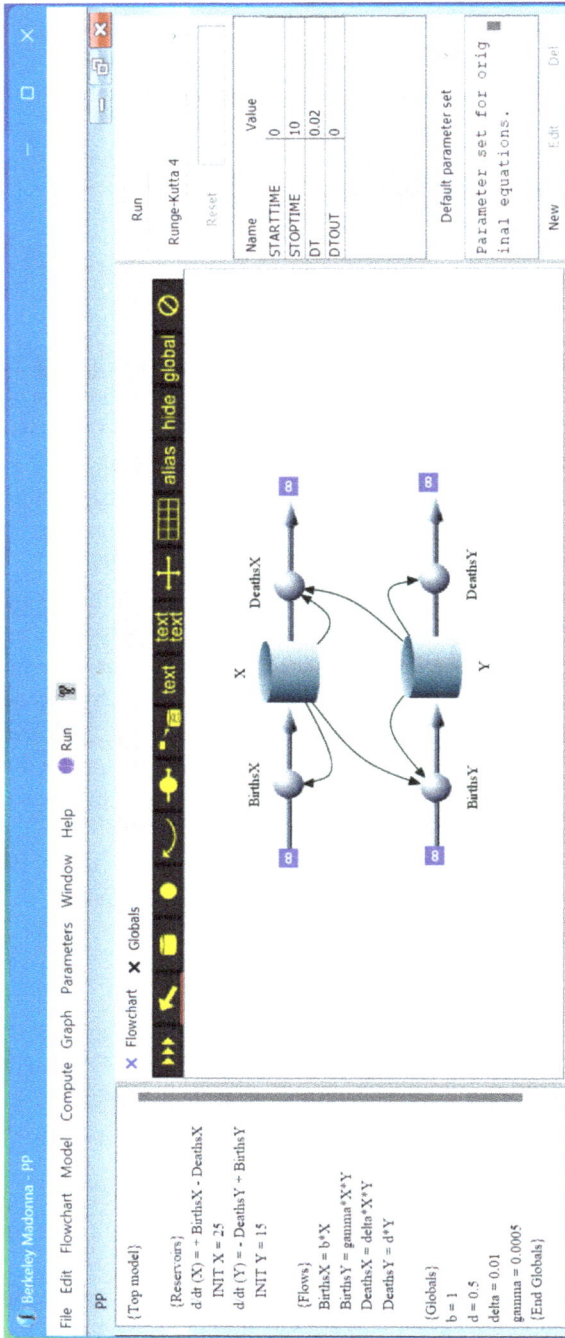

Figure 2.20: The completed flowchart and model for the predator–prey system.

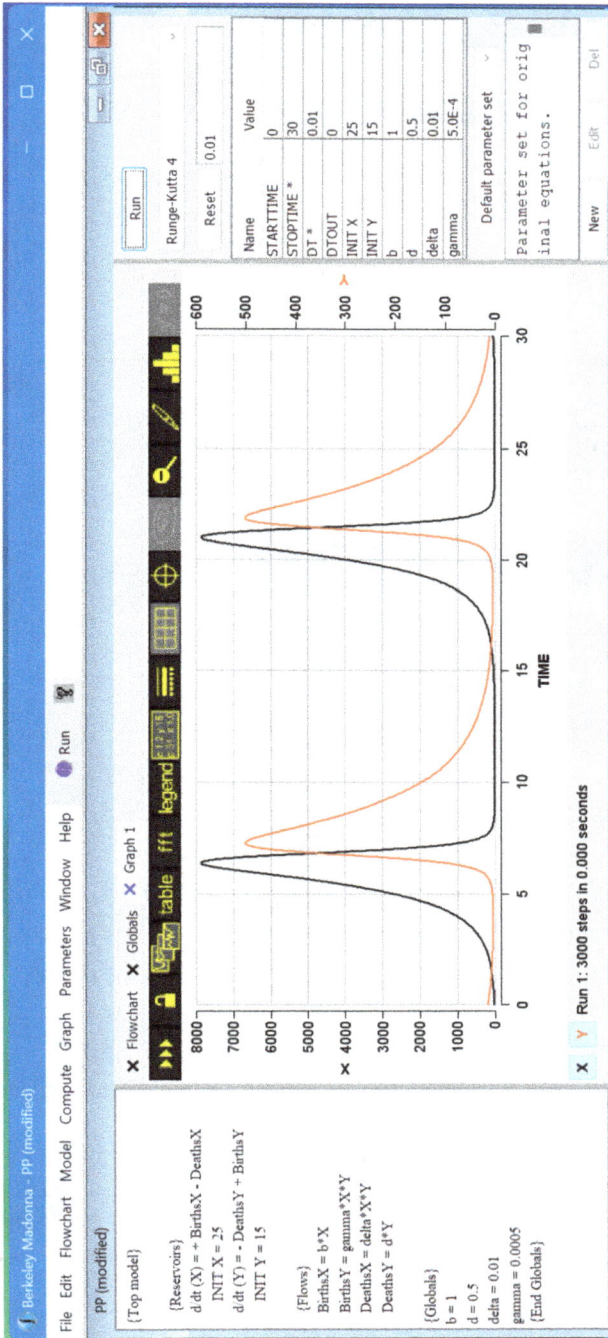

Figure 2.21: Solution curves for the predator–prey model with STOPTIME = 30 and DT = 0.01.

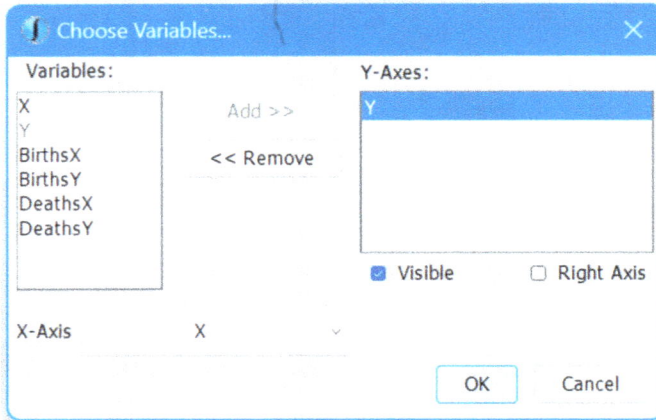

Figure 2.22: Choosing variables for a new graph.

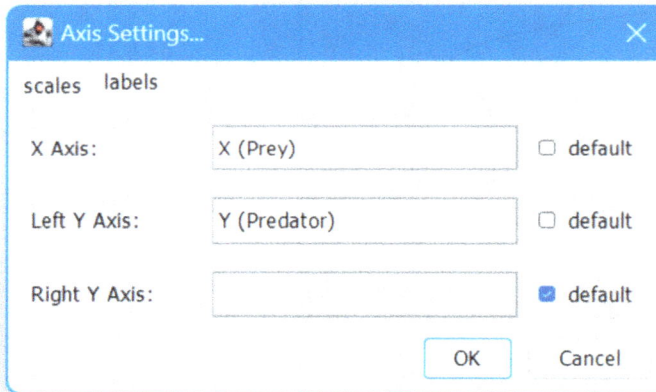

Figure 2.23: Customizing axes labels. Axes scales can customized by clicking on the `scales` tab and then making the desired changes.

mouse was clicked; the coordinates of this point appear in the lower-left corner of Figure 2.25.

Just as shown for the previous SIR example, sliders can be attached to either of the two graphs constructed in Figures 2.21 and 2.24. Be sure to deselect the `Overlay Plots` and `Initial Conditions` buttons before using sliders—see what happens if this is not done. These sliders can then be used to explore the effects of playing around with parameter values of interest.

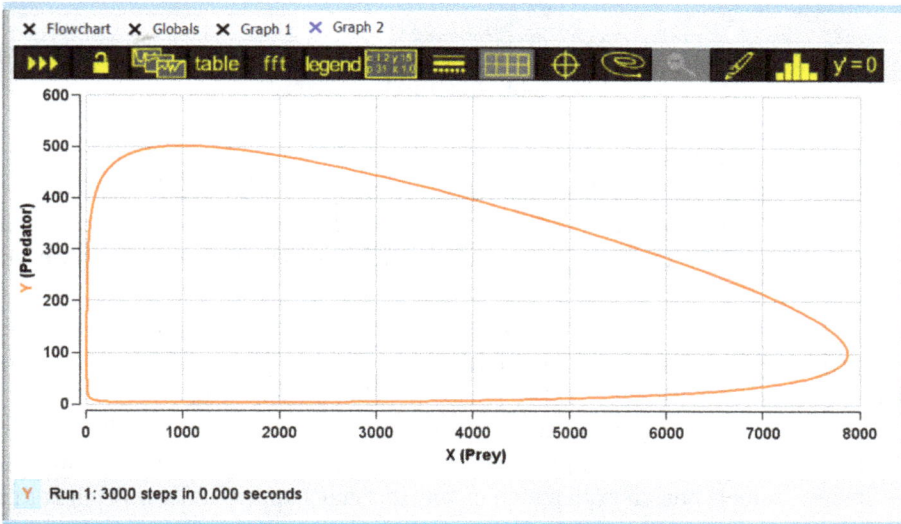

Figure 2.24: Plot of predator population size against prey population size.

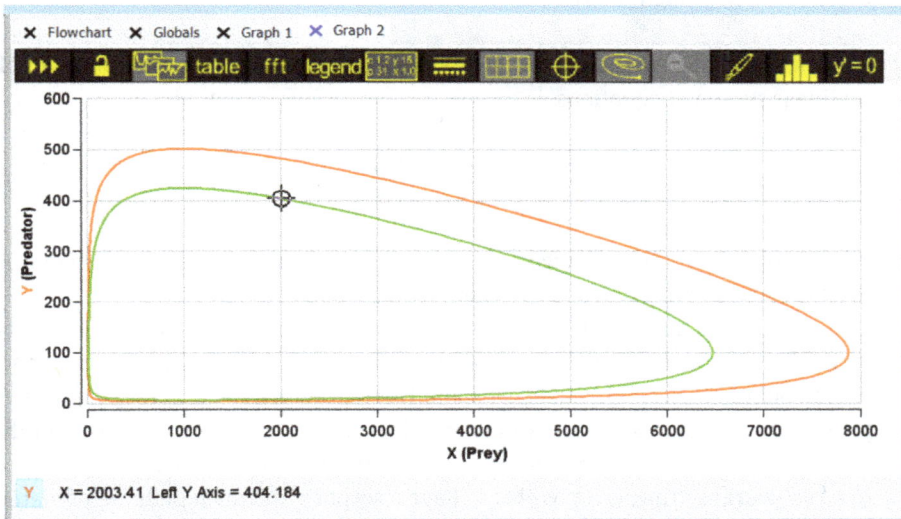

Figure 2.25: Overlaying *XY*-plots of curves with different initial conditions.

2.5.4 Exploratory exercises

As with the previous SIR example, the panel on the left in Figure 2.21 gives details of the system being modeled. This corresponds to the general form

$$\frac{dX}{dt} = bX - \delta XY,$$

$$\frac{dY}{dt} = -dY + \gamma XY,$$

(2.4)

with initial conditions being denoted by $X(0) = X_0$ and $Y(0) = Y_0$. This system of equations, called the *Lotka–Volterra equations*, has long served as a popular first example of modeling predator–prey systems. Observe that the units for the parameters b and d are t^{-1}, and the units for δ and γ are $(\text{counts} \times t)^{-1}$. Here are some exploratory exercises.

Focusing on Graph 2, start by leaving the parameter values as originally defined. Now play around with predator–prey initial population sizes, as was done to produce Figure 2.25, *but* have *only* the Initial Conditions button activated:

1. Does it appear to be possible to have *nonzero* initial conditions that will result in the either of the predator or prey populations dying out? If yes, provide an (approximate) example and an explanation of why this may happen. If no, why might this be?

2. Does it appear possible to have nonzero initial conditions that will guarantee that both the predator and the prey populations will remain constant? If yes, provide these. If no, why might this be?

Notice from System (2.4) that a yes for the previous exercise corresponds to solutions of the nonlinear system of equations

$$X_0(b - \delta Y_0) = 0,$$

$$Y_0(-d + \gamma X_0) = 0.$$

(2.5)

3. Does it appear possible to have nonzero initial conditions that will result in one of the populations dying out, and the other growing nonstop? Explain.

4. Observe that the trivial solution $(0, 0)$ is one solution for the System (2.5). Find formulas for the nonzero initial conditions that guarantee that both the predator and prey population sizes will remain constant. This pair is referred to as the *equilibrium point* of the system. Using the parameter values for the original equations, how do the initial conditions here compare with your answer from Exercise 2?

5. In a real-world setting, is it possible to have a scenario in which both the predator and prey population sizes will remain constant? Why or why not?

Before continuing, reset the initial conditions to the original values ($X_0 = 25$ and $Y_0 = 15$), open Graph 1, and then insert sliders with suitable ranges for the parameters δ and γ for this graph (refer back to Section 2.4.4).

6. With $\gamma = 0.0005$, use the slider for δ to explore how changes to the value of δ influence the solution curves. What happens? What might be a real-world reason for changes in the value of δ?

7. With $\delta = 0.01$, use the slider for γ to explore how changes to the value of γ influence the solution curves. What happens? What might be a real-world reason for changes in the value of γ?

Before continuing, reset the values of δ and γ to their original values ($\delta = 0.01$ and $\gamma = 0.0005$) and, again in Graph 1, remove the sliders for δ and γ and then insert sliders for the parameters b and d with suitable ranges.

8. With $d = 0.5$, use the slider for b to explore how changes to the value of b influence the solution curves. What happens? What might be a real-world reason for changes in the value of b?
9. With $b = 1$, use the slider for d to explore how changes to the value of d influence the solution curves. What happens? What might be a real-world reason for changes in the value of d?

2.5.5 Extensions and other ideas

The Lotka–Volterra model is useful as both an introduction to modeling predator–prey systems, as well as for getting a quick feel for using Berkeley Madonna for population ecology modeling. However, as is the case with all mathematical models involving Nature, the idealized system being modeled is almost guaranteed to be nowhere near to the reality on the ground. But it is possible extend simple models to account for more "stuff."

The predator–prey model presented here may be modified in certain ways to account for additional characteristics of the system being modeled, or even adapted to a different scenario. As previously mentioned, it is always helpful to pose questions before making changes. For example:

− Should any changes to the number of compartments be made?
− How many compartments will there be in the proposed alternative or extended model? Will there be fewer or more?
− Suppose there is an increase in the number of compartments. Do the additional compartments represent distinct populations or distinct sub-populations?
− What are expected roles that these compartments would play?
− Will flows between any compartments exist? If yes, what will the direction of flow between these compartments look like?
− Will the number of parameters involved change from number in the previous predator–prey model? If yes, will there be fewer or more? What will these be?
− How will the compartments and parameters contribute to the definitions of flows out of and/or into compartments (think arcs)?
− What might be useful/informative solution curve plots?

Again, once a model is constructed there are other questions that can be asked. For example:

- For which parameters should sliders be included? Why?
- What information does varying each parameter provide?
- What is a meaningful goal of performing an analysis of the model through parameter variations?

Accepting the "instantaneous mixing" assumption, here are some ways in which the previous predator–prey model might be enhanced, or adapted to a different scenario.
- **Density-dependence**: It is possible to introduce density-dependence on the rate at which the prey population size grows (or decreases) by incorporating a carrying capacity to the ecosystem with respect to the prey. The idea behind this is to have the prey population size start to decrease once the number of prey goes above the carrying capacity. There are a variety of approaches to do this, a popular one is the *logistic growth model*. Would this involve a major change to the previous predator–prey model? If yes, how? If no, what would the change be?
- **Age-dependence**: It could be asked whether age (of the prey and/or the predator) might play a role in the effectiveness of the predator pursuing and catching a prey. How might this idea be incorporated into the previously seen predator–prey model?
- **A two competing-species model**: Moving away from the previous predator–prey model, consider two species that compete with each other for a common food-source. Suppose further that predators of these species are absent from the picture. How might such a scenario be modeled?
- **A competitor for the prey's food source**: Moving back to the previous predator–prey.model, it is possible that the system contains a second species that competes with the prey for the prey's food source. Assuming this (additional) species is not preyed upon by the predator, how might this scenario be modeled?
- **A Competing predator species**: It is also possible that the system contains a second predator species that competes for the prey. Assuming there is no other prey species, how might this scenario be modeled?

There are many other questions that can be asked, for example:
- How can the loss of habitat be incorporated into wildlife population modeling involving compartmental models?
- How can human encroachment be incorporated?
- How can invasive species be brought into the picture?

For purposes of this book, the key to modeling any given population ecology scenario that meets the general form of the first-order system of differential equations (2.1) lies in acquiring a good understanding of the necessary biological and/or ecological background information and then putting together a flowchart in Berkeley Madonna. See Section 2.6.2 for a selection of resources for this.

2.6 Selected resources

The examples presented in Sections 2.4 and 2.5 were chosen for their relative simplicity so as to provide illustrations of constructing flowcharts in Berkeley Madonna for two versions of compartmental models. As demonstrated, the solution of the resulting models and performing a range of simulations is made very straightforward by Berkeley Madonna.

It is anticipated that readers may be interested in delving deeper into compartmental models and Berkeley Madonna to explore more complex applications. In addition to suggested resources for Berkeley Madonna, this section provides a further selection of resources that focus on the background theory through books as well as ideas for possible student or faculty research projects through articles. The bibliographies of the resources listed also contain extensive lists of further potentially useful/interesting applications including, for example, epidemiology, medicine, wildlife ecology, and much more. Though the level at which the content of the resources are delivered do vary quite a bit, they all provide useful background information as well as seeds for interesting basic or advanced research projects.

2.6.1 Berkeley Madonna

The *User Guide* for Berkeley Madonna is available on the Berkeley Madonna website at

https://berkeley-madonna.myshopify.com/pages/technical-faq

For those interested, examples of using Berkeley Madonna in more complex settings can be found in the tutorials [84] and [108], and a list of textbooks that use Berkeley Madonna in a classroom setting can be found on the Berkeley Madonna website at

https://berkeley-madonna.myshopify.com/pages/links

Also, on this site, are links to a Forum site to ask questions and get tips and advice from other users, as well as a *Berkeley Madonna Users Group* at LinkedIn.

2.6.2 Background and theory

It is worth mentioning here that the use of the descriptor "compartmental model" is not necessarily universal, it appears most often in the fields of medicine, epidemiology and ecology. Authors typically refer to compartmental analysis in association with systems of first-order differential equations along the lines of the first-order system of differential equations (2.1). While the number of books that contain "compartmental models" in their title appears to be limited, there are quite a few that cover applications of such models under broader titles.

Here is a partial list of publications, in chronological order, along with brief descriptions of the contents. The level at which the content is presented varies considerably. However, the works listed do contain useful background information and ideas for interesting exercises, which could be worked with Berkeley Madonna.

Compartmental models and their application (1983) [45]
Brief description: It is mentioned in the preface to this book that at the time of its publication in 1983 there were only two existing books that focused on covering both the theory and applications of compartmental models—the first was published in 1969, [9], and the other in 1972, [66]. The author presents the mathematical content informally for readers new to the field. There are two chapters that may be of preliminary interest to those not desiring a detailed treatment of the mathematical theory that justifies many of the applications. Chapter 1 introduces compartmental models, and Chapter 11 discusses applications to biomedical systems, pharmacokinetics, and ecosystems.

Frontiers in mathematical biology (1994) [80]
Brief description: This work is included as it contains a broad collection of articles, some of which may be of interest or of potential use. The content does not specifically address compartmental models, but it does cover a wide range of fields and may provide useful disciplinary background information. Areas covered include cell and molecular biology, organismal biology, evolutionary biology, population ecology, community and ecosystem ecology, and applied biology.

Mathematical models in biology (2005) [39]
Brief description: Written for advanced undergraduates, Part II of this text presents an introduction to continuous models and their applications to population dynamics. Applications to population dynamics discussed include modifications of the Lotka–Volterra equations that incorporate density dependence for the prey, attack capacity of the predator, a competing-species model, and models for multiple species communities. In the area of infectious diseases, discussions on the SIS, SIR, and SIRS are presented with an emphasis on the population biology associated with these models.

Theoretical neuroscience: computational and mathematical modeling of neural systems (2005) [34]
Brief description: This explores the use of theoretical analysis and computational modeling in neuroscience to understand the functions and operations of nervous systems through descriptive, mechanistic, and interpretive models bridging various levels of scientific inquiry.

Mathematical biology I: an introduction (2007) [94]

Brief description: This book is considered a classic introduction to mathematical biology. This first volume by Murray on mathematical biology presents an introduction to the field and the mathematics mainly involves ordinary differential equations that are suitable for undergraduate and graduate courses at different levels. It may be considered a very useful resource for background theory addressing population and infectious disease models, and much more.

Mathematical epidemiology (2008) [25]

Brief description: This collection of lecture notes targeting graduate students covers a wide range of topics in the realm of infectious diseases and compartmental models, including case studies of a range of diseases and a list of suggested exercises and projects. It also contains a chapter devoted to the basic reproduction number; see also [115].

A short history of mathematical population dynamics (2011) [10]

Brief description: This book provides historical background information as well as the mathematical analysis of a selection of classical mathematical models in population biology with complementary reading for undergraduate and graduate students studying mathematical biology. While some of the methods may not involve compartmental models, this may be an interesting resource for background information.

Modeling of living systems: from cell to ecosystem (2012) [97]

Brief description: This book introduces modeling in the life sciences using the "from practice to theory" approach with methods being illustrated by many examples. An historical and general introduction informs the reader how mathematics and formal tools are used to solve biological problems at all levels of the organization of life. The core of this book explains how this is done based on practical examples coming, for the most part, from the author's personal experience. In most cases, data are included so that the reader can follow the reasoning process and even reproduce results presented in the book. The final chapter is devoted to essential concepts and current developments. The main mathematical tools are presented in an Appendix to the book and are written in an adapted language readable by scientists, professionals, or students, with a basic knowledge of mathematics.

Compartmental modeling in the analysis of biological systems (2012) [14]

Abstract: Compartmental models are composed of sets of interconnected mixing chambers or stirred tanks. Each component of the system is considered to be homogeneous, instantly mixed, with uniform concentration. The state variables are concentrations or molar amounts of chemical species. Chemical reactions, transmembrane transport, and binding processes, determined in reality by electrochemical driving forces and constrained by thermodynamic laws, are generally treated using first-order rate equations.

This fundamental simplicity makes them easy to compute since ordinary differential equations (ODEs) are readily solved numerically and often analytically. While compartmental systems have a reputation for being merely descriptive they can be developed to levels providing realistic mechanistic features through refining the kinetics. Generally, one is considering multicompartmental systems for realistic modeling. Compartments can be used as "black box" operators without explicit internal structure, but in pharmacokinetics compartments are considered as homogeneous pools of particular solutes, with inputs and outputs defined as flows or solute fluxes, and transformations expressed as rate equations.

Notes: While this article focuses on pharmacokinetics, it may serve as a useful resource for ideas on how to tackle compartmental models for more complex settings that the examples presented in this chapter may extend to.

Methods and models in mathematical biology: deterministic and stochastic approaches (2015) [92]
Brief description: This advanced text covers an introduction to compartmental modeling in the discrete and continuous setting through applications to ecology, epidemiology, reaction kinetics, neuronal activity, and evolution.

Mathematical models in epidemiology (2019) [24]
Brief description: This comprehensive and self-contained work presents an introduction to the mathematical modeling and analysis of disease transmission. It provides an introduction to simple compartmental models for infectious diseases, addresses models for endemic diseases and epidemics, and provides discussions of models for some specific diseases. It also discusses some advanced concepts, including introducing age and spatial structure to a model and incorporating mobility, behavior, and time scales.

Elementary differential equations and boundary value problems (2021) [23]
Brief description: This latest edition of the popular undergraduate text with the same name contains two chapters (7 and 9) that address areas that might be interesting to those looking for relevant mathematical background.

Introducing mathematical biology (2023) [17]
Brief description: This open-source text covers a range of topics that involve compartmental modeling. It targets undergraduate and graduate mathematics students who have not studied mathematical biology before, life-sciences students with an interest in modeling, and high school students interested in university-level material on mathematical biology. Some mathematical knowledge is assumed, and the mathematical models used are all expressed as systems of first-order ordinary differential equations. Areas covered include population ecology, infectious diseases, immune and cell dynamics, gene networks, and pharmacokinetics.

Epidemiology of infectious diseases: A human view (2023) [42]
Brief description: The following description is provided by the publisher.

> Current textbooks provide a strong biomedical view on epidemics. In this textbook, the biomedical view will be extended to a human view including insights from humanities, social sciences. This extension challenges us all the more to combine the requirement of scientific objectivity with the subjectivity inherent to human life. In addition, the biomedical view is deepened using knowledge of botanical epidemiology with respect to 'evolutionary dynamics of pathogens' and 'epidemic spread of pathogens'. Biomedical oriented students and senior scientists are invited to reflect on the multidimensional, subjective, character of epidemics. Reflections that may enable appropriate, human, management of epidemics.

This may contain ideas that could be useful in extending models beyond the "hard sciences" point of view.

2.6.3 Research and ideas

The following selection of articles, also presented in chronological order, includes articles of potential use to students, teachers, and researchers. Some of the articles contain useful background theory, and others cover a range of applications. For each article listed, the abstract is provided for general informational purposes. This is followed by brief notes on how the article might serve as a useful resource on background information/theory, contribute to the understanding of a similar proposed project, or provide useful ideas on tackling a seemingly different, but actually equivalent project.

Analysis of linear compartment models for ecosystems (1974) [91]
Abstract: This paper discusses the mathematical validity of the class of linear periodically time varying compartment models, which are commonly employed for the analysis of ecosystems. Valid mathematical models for such systems must have unique stable periodic solutions to which all other solutions converge. In addition solutions of such ecosystem models which are initially positive must remain positive. The results presented in this paper provide a mathematical verification for the application of linear donor controlled compartment models to ecosystem analysis.

Notes: This article discusses linear systems, as opposed to nonlinear systems. So, it may provide ideas for ecosystems that lend themselves to first-order linear systems of differential equations.

Population dynamics: an introduction to differential equations (1975) [26]
Abstract: In this paper, a number of population models, which lead to differential equations, are derived. First-order variables separable equations are formulated from the Malthusian population model and its extension to include crowding effects. Age structure effects are shown to lead to differential equations with a time lag and the dynamics

of exploited fish populations are briefly examined. Two models of interacting species are examined, predator–prey and competing species, both of which lead to simultaneous coupled nonlinear differential equations but with solutions, which have vastly differing characteristics.

Notes: This article is a useful brief resource for a variety population dynamics models.

On the mathematical foundations of compartmental analysis in biology, medicine, and ecology (1978) [105]

Abstract: The basic equations of compartmental analysis are a system S of differential equations, which govern the exchanges of material among various compartments and an environment. The equations are ordinarily nonlinear. In this paper, we show that if certain natural conditions are satisfied S possesses an equilibrium point, and that if additional reasonable conditions are met the equilibrium point is unique and globally stable. These results have natural interpretations and, in particular, provide an analytical basis for the use of the familiar linear tracer-analysis equations. We also consider the case in which S takes into account cyclic variations with a given period τ (this case often arises in ecological studies) and show that under certain reasonable conditions S has a τ-periodic solution that is approached by every solution of S.

Notes: This article may provide insights into equilibrium points, mentioned previously in the exploratory exercises for the predator–prey example.

Modeling projects in a differential equations course (1998) [31]

Abstract: Much has been said about the benefits of using projects in teaching the calculus sequence. Here, we discuss the value of student designed, in-depth, modeling projects in a differential equations course and how to prepare them for it.

Notes: As the abstract suggests, some teachers may find this article has useful ideas for their own classrooms.

Global analysis of a deterministic and stochastic nonlinear SIRS epidemic model (2011) [79]

Abstract: We present in this paper an SIRS epidemic model with saturated incidence rate and disease-inflicted mortality. The global stability of the endemic equilibrium state is proved by constructing a Lyapunov function. For the stochastic version, the global existence and positivity of the solution is shown, and the global stability in probability and pth moment of the system is proved under suitable conditions on the intensity of the white noise perturbation.

Notes: This article may find use in providing ideas and background theory appropriate for extending the previously encountered SIR model.

Compartmental models of migratory dynamics (2011) [74]

Abstract: Compartmentalization is a general principle in biological systems, which is observable on all size scales, ranging from organelles inside of cells, cells in histology, and up to the level of groups, herds, swarms, meta-populations, and populations. Compartmental models are often used to model such phenomena, but such models can be both highly nonlinear and difficult to work with.

Fortunately, there are many significant biological systems that are amenable to linear compartmental models, which are often more mathematically accessible. Moreover, the biology and mathematics is often so intertwined in such models that one can be used to better understand the other. Indeed, as we demonstrate in this paper, linear compartmental models of migratory dynamics can be used as an exciting and interactive means of introducing sophisticated mathematics, and conversely, the associated mathematics can be used to demonstrate important biological properties not only of seasonal migrations but also of compartmental models in general.

We have found this approach to be of great value in introducing derivatives, integrals, and the fundamental theorem of calculus. Additionally, these models are appropriate as applications in a differential equations course, and they can also be used to illustrate important ideas in probability and statistics, such as the Poisson distribution.

Notes: The contents of this article may be of interest to instructors wishing to illustrate the connections between biological systems and mathematics. Also, the article may provide useful ideas associated with modeling migratory dynamics.

Alfred J. Lotka and the origins of theoretical population ecology (2015) [72]

Brief description: This article provides an outline of the origins of population ecology through the works of Lotka and Volterra.

Notes: This article is included because it provides an interesting short history of Lotka and Volterra, and their contributions to modeling in population ecology.

An evolutionary computing approach for parameter estimation investigation of a model for cholera (2015) [7]

Abstract: We consider the problem of using time-series data to inform a corresponding deterministic model and introduce the concept of genetic algorithms (GA) as a tool for parameter estimation, providing instructions for an implementation of the method that does not require access to special toolboxes or software. We give as an example a model for cholera, a disease for which there is much mechanistic uncertainty in the literature. We use GA to find parameter sets using available time-series data from the introduction of cholera in Haiti and we discuss the value of comparing multiple parameter sets with similar performances in describing the data.

Notes: Background information for modeling a cholera epidemic can be obtained from this article, and findings included may prove useful in preparing a compartmental model of cholera.

Modeling the fear effect in predator–prey interactions (2016) [118]

Abstract: A recent field manipulation on a terrestrial vertebrate showed that the fear of predators alone altered anti-predator defenses to such an extent that it greatly reduced the reproduction of prey. Because fear can evidently affect the populations of terrestrial vertebrates, we proposed a predator–prey model incorporating the cost of fear into prey reproduction. Our mathematical analyses show that high levels of fear (or equivalently strong anti-predator responses) can stabilize the predator–prey system by excluding the existence of periodic solutions. However, relatively low levels of fear can induce multiple limit cycles via subcritical Hopf bifurcations, leading to a bi-stability phenomenon. Compared to classic predator–prey models, which ignore the cost of fear where Hopf bifurcations are typically supercritical, Hopf bifurcations in our model can be both supercritical and subcritical by choosing different sets of parameters. We conducted numerical simulations to explore the relationships between fear effects and other biologically related parameters (e. g., birth/death rate of adult prey), which further demonstrate the impact that fear can have in predator–prey interactions. For example, we found that under the conditions of a Hopf bifurcation, an increase in the level of fear may alter the direction of Hopf bifurcation from supercritical to subcritical when the birth rate of prey increases accordingly. Our simulations also show that the prey is less sensitive in perceiving predation risk with increasing birth rate of prey or increasing death rate of predators, but demonstrate that animals will mount stronger anti-predator defences as the attack rate of predators increases.

Notes: The mathematics involved in this article (bifurcations in particular) may be a little advanced. However, incorporating a fear factor in a model is an interesting and realistic idea.

A compartmental model of animal behavior (2016) [33]

Abstract: Animal behavior is integral to fitness and arises from complex interactions between internal and external factors. An understanding of how external environmental factors drive animal behavior is important for understanding the way organisms adapt to environmental perturbations such as climate change. Glaucous-winged gulls (*Larus glaucescens*) at Protection Island, Strait of Juan de Fuca, Washington display a variety of behaviors on the colony during the breeding season. The most common gull behaviors are sleeping, preening, and resting. I used a system of four differential equations to predict numbers of sleeping, preening, and resting gulls on the colony as a function of seven environmental factors: hour of day, tide height, solar elevation, heat index, humidity, wind speed on the colony, and wind speed over open water. The model explained 65 %, 51 %, 44 %, and 32 % of the variability in colony attendance, sleep, preen, and rest dynamics, respectively. Similarly, model validation on an independent data set predicted 70 %, 64 %, 60 %, and 47 % of the variability in colony attendance, sleep, preen, and rest dynamics, respectively.

Notes: The ideas in this undergraduate honors thesis may be of direct or indirect interest because they can be used to describe a different (but analogous) "story."

Interdisciplinary education—a predator–prey model for developing a skill set in mathematics, biology, and technology (2017) [116]

Abstract: The science of biology has been transforming dramatically and so the need for a stronger mathematical background for biology students has increased. Biological students reaching the senior or post-graduate level often come to realize that their mathematical background is insufficient. Similarly, students in a mathematics programme, interested in biological phenomena, find it difficult to master the complex systems encountered in biology. In short, the biologists do not have enough mathematics and the mathematicians are not being taught enough biology. The need for interdisciplinary curricula that includes disciplines such as biology, physical science, and mathematics is widely recognized, but has not been widely implemented. In this paper, it is suggested that students develop a skill set of ecology, mathematics, and technology to encourage working across disciplinary boundaries. To illustrate such a skill set a predator–prey model that contains self-limiting factors for both predator and prey is suggested. The general idea of dynamics is introduced and students are encouraged to discover the applicability of this approach to more complex biological systems. The level of mathematics and technology required is not advanced; therefore, it is ideal for inclusion in a senior-level or introductory graduate-level course for students interested in mathematical biology.

Notes: Like the previously mentioned article [31], instructors may find this article useful for ideas on communicating foundational ideas in biomathematical modeling.

An introduction to compartmental modeling for the budding infectious disease modeler (2018) [19]

Abstract: Mathematical models are ubiquitous in the study of the transmission dynamics of infectious diseases. In particular, the classic "susceptible-infectious-recovered" (SIR) paradigm provides a modeling framework that can be adapted to describe the core transmission dynamics of a range of human and wildlife diseases. These models provide an important tool for uncovering the mechanisms generating observed disease dynamics, evaluating potential control strategies, and predicting future outbreaks. With ongoing advances in computational tools as well as access to disease incidence data, the use of such models continues to increase. Here, we provide a basic introduction to disease modeling that is primarily intended for individuals who are new to developing SIR-type models. In particular, we highlight several common issues encountered when structuring and analyzing these models.

Notes: This article provides a nice introduction to using compartmental models to model infectious diseases, and extend or adapt known models for different scenarios.

Using harvesting models to teach modeling with differential equations (2019) [37]

Abstract: Harvesting models based on ordinary differential equations are commonly used in the fishery industry and wildlife management to model the evolution of a population depleted by harvest mortality. We present a project consisting of a series of scenarios based on fishery harvesting models to teach the application of theoretical concepts learned in a differential equations course to scenarios encountered in real fisheries. These projects require a thorough understanding of simplifying assumptions inherent in various models, as well as a qualitative analysis of phase portraits, bifurcations, and stability of steady states. Parameters are estimated and equations are sometimes solved both analytically and numerically. Students learn to respond to a professional request from a fishery in the form of a scientific report, which requires organizing and communicating assumptions, models, solution methods, results, and a final recommendation with clarity and professionalism.

Notes: Here is another article that instructors might find useful for teaching ideas. While some of the analyses' tools may be a little advanced, the emphasis on acquiring a thorough understanding of simplifying assumptions will be valuable.

A mathematical model for the effect of social distancing on the spread of COVID-19 (2020) [109]

Abstract: Social distancing is an effective method of impeding the spread of a novel disease such as severe acute respiratory syndrome coronavirus 2 (SARS-CoV-2), but is dependent on public involvement and is susceptible to failure when sectors of the population fail to participate. A standard SIR model is largely incapable of modeling differences in a population due to the broad generalizations it makes such as uniform mixing and homogeneity of hosts, which results in lost detail and accuracy when modeling heterogeneous populations. By further compartmentalizing an SIR model, via the separation of people within susceptible and infected groups, we can more accurately model epidemic dynamics and predict the eventual outcome, highlighting the importance of societal participation in social distancing measures during novel outbreaks.

Notes: This article provides a nice illustration of how a known model (like the SIR model) can be extended to incorporate differences in a population though additional compartmentalizing.

The SEIRS model for infectious disease dynamics (2020) [18]

Brief description: This short article discusses the inclusion of features to the basic SIR model that account births, deaths, latency, and lost immunity.

Mathematical modeling, analysis, and simulation of the COVID-19 pandemic with behavioral patterns and group mixing (2021) [95]

Abstract: Due to the rise of COVID-19 cases, many mathematical models have been developed to study the disease dynamics of the virus. However, despite its role in the spread of COVID-19, many SEIR models neglect to account for human behavior. In this project, we develop a novel mathematical modeling framework for studying the impact of mixing patterns and social behavior on the spread of COVID-19. Specifically, we consider two groups, one exhibiting normal behavior who do not reduce their contacts and another exhibiting altered behavior who reduce their contacts by practicing non-pharmaceutical interventions such as social distancing and self-isolation. The dynamics of these two groups are modeled through a coupled system of ordinary differential equations that incorporate mixing patterns of individuals from these groups, such that contact rates depend on behavioral patterns adopted across the population. Additionally, we derive the basic reproduction number, perform numerical simulations, and create an interactive dashboard.

Notes: Like the previous article, this article presents ideas that may find use in other scenarios.

Mathematical modeling using scenarios, case studies, and projects in early undergraduate classes (2023) [43]

Abstract: Mathematical modeling has great potential to motivate students towards studying mathematics. This article discusses several different approaches to integrating research work with a second-year undergraduate, mathematical modeling subject. I found sourcing papers from the areas of epidemiology and ecology to be a fruitful source area, particularly models involving only two or three coupled differential equations. These models were amenable to students as well as interesting and relevant to students because they came from real research papers. I will describe the use of scenarios and case studies in lectures, and group projects for assessment. The scenarios and case studies were published in a textbook that I wrote. Scenarios, case studies and projects provided an opportunity to expose students to some novel applications of differential equations. One example is developed here as a classroom note: modeling the dynastic cycles in Chinese history.

Notes: This article will be of interest to instructors who wish to use a case studies or a project-based approach to teaching biomathematical modeling.

3 Agent-based modeling

3.1 Introduction

This chapter expands on the presentation in [5] with illustrations of an alternative approach to modeling the scenarios described in the examples of Chapter 2. The software used here is *NetLogo*, [121], a free open-source software package that is designed specifically for building and exploring *agent-based models* (typically abbreviated as ABM). NetLogo was developed by Uri Wilensky in 1999, and as described on its website,

> NetLogo is a multi-agent programmable modeling environment. It is used by many hundreds of thousands of students, teachers, and researchers worldwide. It also powers HubNet participatory simulations.

The features available in NetLogo lend themselves to supporting basic and advanced research projects, making it well suited for beginners in the area of agent-based modeling. Following the spirit of this book, this chapter focuses on an introductory exploration of agent-based models for the curious teacher, student, or researcher who is unfamiliar with agent-based modeling, but is interested in an informative but brief exposure to the field through NetLogo.

3.2 Agent-based models in general

The term *individual-based model* (typically abbreviated as IBM) also appears in the literature. It turns out that the terms agent-based modeling and individual-based modeling are often used interchangeably and the protocol for designing and implementing such models is the same; see, for example, [36], [50], or [117]. In this chapter, only the term agent-based model (or modeling) is used.

Providing a concise mathematical representation of an agent-based model, as for a compartmental model, is at best very difficult (if possible at all). However, it is possible to describe strengths of agent-based modeling by touching on what it entails.

It is evident from the literature that agent-based modeling stands out as a powerful computational technique used across a wide range of disciplines to simulate complex systems comprised of autonomous *agents*. These agents, which could represent individuals, organizations, or even simple entities such as cells or molecules, operate within a defined environment and follow sets of rules governing their characteristics/attributes, behavior, and interactions. Agent-based modeling offers a bottom-up approach to modeling, where emergent phenomena arise from the interactions of individual agents with each other and/or their environment rather than being explicitly programmed into the model. The concept of agents in agent-based models draws inspiration from observations in social sciences, biology, and other fields where individual entities exhibit distinct behaviors that collectively shape the dynamics of the system as a whole. Agent-

https://doi.org/10.1515/9783111609560-003

based modeling provides a flexible framework for studying phenomena ranging from the behavior of financial markets to the spread of infectious diseases, ecological system dynamics, urban dynamics, and much more.

In agent-based models, agents typically possess attributes, exhibit behaviors, and are subject to rules that govern their interactions with each other and their environments. These interactions often lead to the emergence of complex patterns, dynamics, and phenomena that would be difficult to predict solely from understanding the behavior of individual agents in isolation. One of the key strengths of agent-based modeling lies in its ability to capture heterogeneity, nonlinearity, and feedback loops inherent in many real-world systems. By representing agents as autonomous entities with diverse characteristics and decision-making processes, agent-based modeling enables researchers to explore how micro-level interactions give rise to macro-level patterns and behaviors.

As a consequence of all of the above, agent-based modeling has found applications in diverse fields including economics, sociology, ecology, epidemiology, transportation planning, and more. Its versatility allows researchers to investigate various "what-if" scenarios, test the effects of different policies or interventions, and gain insights into the underlying mechanisms driving complex systems.

3.3 The ODD protocol

An important and very helpful protocol for preparing an agent-based model was compiled by Grimm et al. and originally published in [50]. This was reviewed and updated first in [51] and later in [54]. Referred to as the *ODD protocol*, this describes the modeling process as involving seven elements grouped into three blocks. These are briefly summarized below—refer to [54] for complete and formal discussions along with supplements containing helpful guiding questions and answers, and examples.

Overview:
Focuses on providing an idea of the overall purpose, structure, and process of a model. This serves as a guide for what is to follow.
- *Purpose*: Typically associated with introductory comments, this involves providing a clear, concise, and complete description of the purpose of the model.
- *Entities, state variables and scales*: This involves the identification of entities to be modeled, and the full set of variables that characterize the entities (such as individuals or habitat units) associated with the model. These should be clearly described and assigned meaningful dimensions and scales of measure.
- *Process overview and scheduling*: This involves compiling a complete description of the environmental and individual processes that are to be built into the model. Preparing a schedule of when and how these occur will help in building *modular code*.

Design concepts:

A checklist of eleven items is provided to aid in establishing a common framework for designing and communicating any given agent-based model. For example, the first three in the list include asking (and answering) questions about: *Basic principles*—Which general concepts, theories, hypotheses, or modeling approaches are underlying the model's design? *Emergence*—What key results or outputs of the model are modeled as emerging from the adaptive traits, or behaviors, of individuals? *Adaptation*—What adaptive traits do the individuals have? What rules do they have for making decisions or changing behavior in response to changes in themselves or their environment?

Details:

This involves bringing together information from the *Overview* and *Design Concepts* blocks, along with any needed technical details to finally construct and implement the model.

- *Initialization*: Establish the initial state of the "world" within, which simulations of the model are to be performed. That is, decide upon parameter values and initial values for all state variables, and reasons for choices made.
- *Input data*: Determine whether the model uses input from external sources or other models to represent processes that produce changes over time.
- *Submodels*: Revisit the *Process overview and scheduling* element, but with further detail. Identify and describe those tasks that can be thought of as *modules* within the broader picture. How these modules tie together to make up the complete model can often be summarized using a flow chart.

The benefits of using this protocol become very evident when modeling complex systems.

For the present, going through two simple examples of agent-based models, such as are provided in Sections 3.5 and 3.7, is a good way to get a feel for what goes on.

3.4 NetLogo preliminaries

As previously mentioned, NetLogo is an open-source, multi-agent, programmable modeling environment that can be used to simulate natural and social phenomena as well as the dynamics of complex systems over time. The user-friendly interface of NetLogo makes it accessible to both novices and experienced users, and its flexibility makes it a versatile tool that can be used across disciplines. Its high-level programming language is specifically designed for simulating and modeling agent behaviors and interactions, and is easy to learn and implement. Another feature of NetLogo is the ease with which sliders can be used to observe immediate changes in simulation results. This enhances its value as a supporting tool for analysis and decision-making, as well as making it an ideal teaching tool for demonstrating dynamical concepts to nontechnical audiences. The link

https://ccl.northwestern.edu/netlogo/bind/

on the NetLogo website provides detailed instructions on downloading and installing the software and much more. Also of particular use on the NetLogo website is the *NetLogo User Manual* at

https://ccl.northwestern.edu/netlogo/docs/

and the *NetLogo Dictionary* at

https://ccl.northwestern.edu/netlogo/docs/dictionary.html,

which contains the complete list of code syntax along with descriptions and examples.

3.4.1 Key features

NetLogo is particularly popular in the fields of complex systems, social sciences, biology, ecology, and the environmental sciences due to its ease of use, flexibility, and powerful modeling capabilities. Some key features that contribute to this popularity include the following:

- **Agent-based modeling environment**. NetLogo provides a user-friendly interface for creating and simulating agent-based models. Users can define different types of agents, called *turtles* (representing individual entities) and *patches* (representing discrete spatial locations), and specify rules governing their behavior and interactions.
- **Turtle graphics**. NetLogo features a built-in graphical display where agents, represented as turtles, move and interact with each other and/or their environment within a two-dimensional grid, referred to as the *World*. Users can customize the appearance of turtles, patches, and other elements of the simulation environment.
- **Programming language**. The programming language and syntax used by NetLogo are easy to learn and implement, and are accessible to users with varying levels of programming experience. The language allows users to define procedures, create variables, and control the behavior of agents and the simulation environment.
- **Definition of agent behaviors**. Users can define the behaviors of individual agents. This includes specifying how agents move, interact with other agents, perceive their environment, and respond to changes in their surroundings.
- **Model library**. NetLogo provides a rich library of prebuilt sample models covering a wide range of topics, including biology, ecology, economics, sociology, and more. These models can serve as templates for users to explore, modify, and extend, making it easier to get started with building models to perform complex simulations.

- **Tools for experimentation and analysis**. NetLogo supports experimentation and analysis tools that allow users to run multiple simulations with varying parameter values, collect data, visualize results, and analyze the behavior of the agents and environment being simulated by model over time. This enables users to test hypotheses, explore different scenarios, and gain insights into the dynamics of complex systems.
- **Community and resources**. NetLogo has a large and active community of users, developers, and educators who contribute to the development of the software and share resources, tutorials, and best practices. This vibrant community provides support and collaboration opportunities for users interested in agent-based modeling; see

https://ccl.northwestern.edu/netlogo/models/community/index.cgi

In summary, NetLogo is a free and well-supported platform that provides a powerful and intuitive tool for building and analyzing agent-based models through simulations. All this makes it an invaluable tool for students, educators, researchers, and practitioners interested in studying complex systems and emergent phenomena.

3.4.2 Key components

Getting started with the process of using NetLogo to do agent-based modeling for the first time is fairly straightforward once the purposes of certain key components are understood. Before beginning, there are three tabs, two drop-down menus, and a button that appear in the opening screen of NetLogo that are worth mentioning; see Figure 3.1.

Interface:
The model interface window is where the simulation is displayed. Also, it is here that buttons, sliders, and switches that control parameter values and the model simulation are placed. Additionally, dynamic graphs and more that can be used to monitor quantitative output associated with the model simulation can also be placed here. The option to insert such features using the Add button and drop-down menu next to it is contained in this window; see Figure 3.1. The blank (black) square region, which will be referred to as the canvas, is where the results of each iteration of the simulation itself are displayed. The canvas can be moved around and/or resized by right-clicking on it and choosing Select and then dragging it or resizing it as desired.

Settings:
A NetLogo canvas has certain default settings that suffice for most purposes, but may be changed as needed; see Figure 3.2. In the default setting, the canvas, named the *World*, is a coordinate plane that consists of a 33×33 grid centered at the origin, $(0, 0)$. This

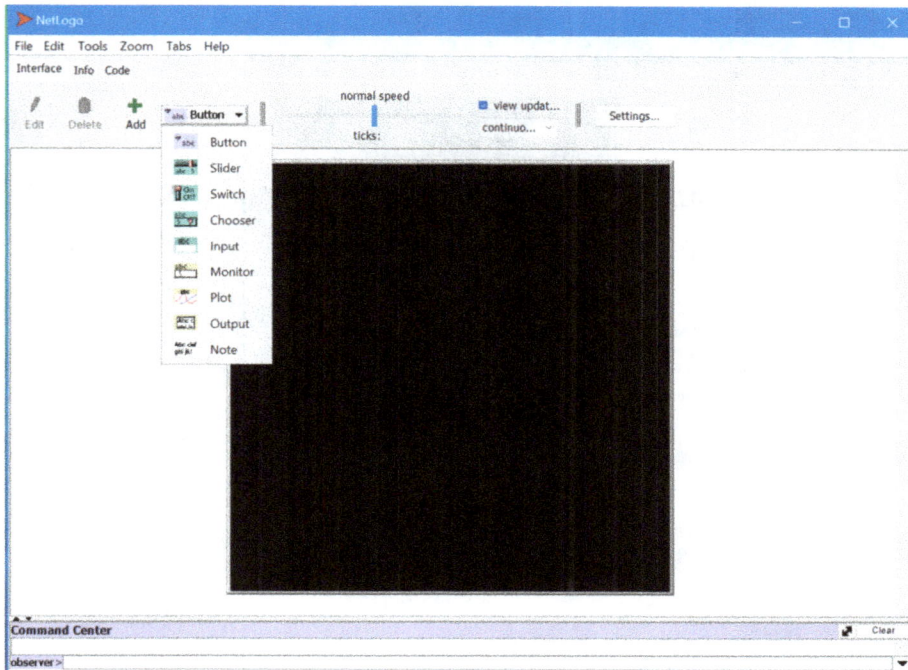

Figure 3.1: Opening view of the NetLogo window. The tabs of interest are Interface (the current view), Info, and Code. The drop-down menu next to the Add button will find use in the examples to follow, but the Command Center at the bottom of this window is better left for when users have achieved a suitable level of comfort with NetLogo. The view updates option is typically left on the continuous setting, and the Settings... button to the right of this opens the pop-up window shown in Figure 3.2.

grid is a planar representation of a torus, which results in a useful feature, the *world wrap*—if an agent travels off the screen on one side, it will reappear on the opposite side. Another setting that is of importance is the *view updates* option; see Figure 3.1. If this is set at continuous, the model will continue updating in between *ticks* (or time steps). This takes up a lot of processing space and may slow down the visual displays of the model. If the setting is on ticks, the model's visual display will only update after each tick. For more complex models, this leads to a more efficient performance, the downside being that the model's visual display will update in steps rather than continuously.

Info:
The Info tab opens the documentation window for the model. Suggested headers along with brief descriptions of what to place under the headers are provided for model creators to fill in useful information about their model, about possible applications, and on possible extensions. If shared with the NetLogo user community on the NetLogo website, this provides potential users of the model with valuable information that enables

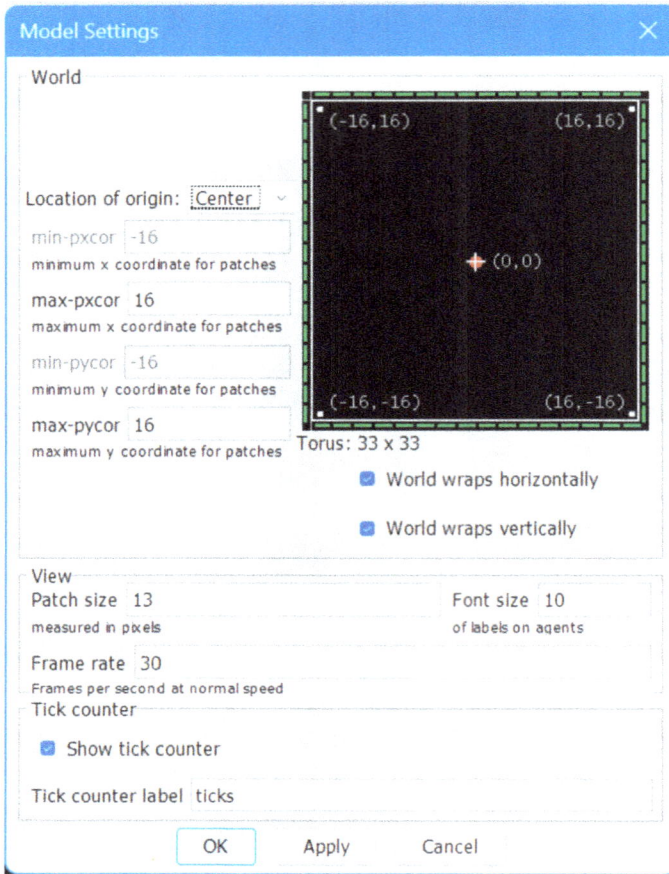

Figure 3.2: The Model Settings window shown here is where the settings are defined for the canvas on which the simulations are displayed. For most simple scenarios, the displayed default settings suffice.

them to build off of others' work efficiently and contribute to the continually growing database of developed agent-based models in NetLogo.

Code:

Code for a NetLogo program is entered in the Code window. The programming language of NetLogo allows for the writing of clear modular code to perform desired tasks. As with other programming languages, *procedures* (or functions) in NetLogo are self-contained sets of instructions for tasks to be performed by the program. Each of these begins with the keyword to followed by the name, and then the body of the procedure (the instructions). The keyword end is used to signal the end of a procedure. Another important component of any NetLogo program is documentation—for the benefit of both the program's developer and future users. Informative comments should be placed wherever helpful *after* a semicolon. Every new comment line must begin with a semicolon. The following example provides a quick introduction to coding an agent-based model in Net-Logo, and this is followed by some tips on how to go about broadening skills in this area.

3.5 An infectious disease model

Consider creating an agent-based model for the generic SIR model described in Section 2.4. That is, suppose one or more individuals who have a nonfatal but very infectious disease enter a population of individuals who have no immunity to the disease. What happens next? A flowchart such as shown in Figure 3.3 can help in visualizing a way to answer to this question.

The task at hand is to prepare an agent-based model that will describe the spread of the disease along the lines of the flowchart shown in Figure 3.3. While not formally outlined, the previously mentioned ODD protocol (see Section 3.3) is informally used here to develop and build the corresponding SIR model. Readers are encouraged to consciously associate parts of the following discussions with components of the ODD protocol described in [54].

3.5.1 Setting the stage

The first steps in constructing an agent-based model for the system in question involve establishing some assumptions and getting a feel for what the variables and parameters should be. These will aid in putting together the various components that describe the mechanics of the system being modeled.

The goal is to track the number of susceptible, infected, and recovered individuals over some time period, $0 \leq t \leq T$ days. For this purpose, assume the following:

– For the time period in question, there are no births or deaths, and neither emigration nor immigration occur. So, the total size of the population remains constant.

– Instantaneous mixing, as for the compartmental model described in Section 2.4, does *not* occur. However, all individuals in the population do move around the community freely and interact with each other in a random manner.

– The process begins with a small number of infected individuals, with the remaining individuals being susceptible to infection. At the start, there are no individuals who have recovered from the disease.

– For every interaction between an infected and susceptible individual, there is a chance (expressed as a percentage) that the susceptible individual will become infected.

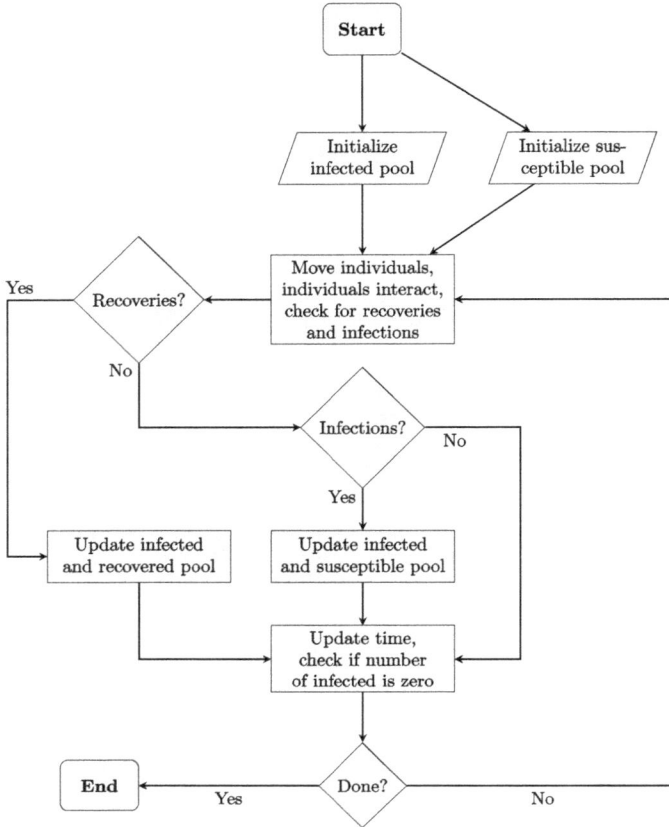

Figure 3.3: A flowchart to describe the spread of an infectious disease under the SIR structure. The looping/simulation stops when there are no infected individuals left.

- Infected individuals immediately becomes infectious and can pass the disease on to susceptible individuals they come in contact with.
- Every infected individual has a chance of recovering from the disease after some individual-dependent time.
- An individual who recovers from the disease acquires immunity, at least for the time period in question.
- At some point in the specified time period $0 \leq t \leq T$ days, it may be that there will be no more infected individuals. This signals the end of the disease.

When preparing to write NetLogo code for an agent-based model, it is helpful to visualize the system at both the individual and communitywide levels. Meaning, deciding on what processes are involved, and which of these can best be placed in a procedure. Having a flowchart such as in Figure 3.3 helps in preparing a plan for constructing code for the model.

3.5.2 Preparing the canvas and the code

For purposes of this exercise (and for most simple applications), the view updates is left as continuous and default model settings for the *World* are used; see Figures 3.1 and 3.2. Coding ideas for this example were acquired and adapted from an introductory model in the curricular unit in the NetLogo Models Library, called *epiDEM* (*Epidemiology: Understanding Disease Dynamics and Emergence through Modeling*), [125].

For this example, the *World* provides a spatial representation of the community in question, and for the scenario in question, it is worth mentioning that individuals move around and may or may not spread the disease but they do not interact with their (spatial) environment and do not cause changes to the physical spaces they pass through.

Having decided on what the variables and parameters should be, it is useful to make a decision on which variables or parameters should be passed into the code through sliders. In this case, for example, it might be preferred that sliders for the cutoff time, the size of the population in question, initial percentage of the population that is infected, the chance a susceptible individual will become infected, the chance an infected individual will recover, and the average recovery time for an infected individual. Syntax used to represent these along with brief descriptions are as follows:

- max-time is used to set the maximum number of days the simulation will run through (see Figure 3.4).
- pop-size is used to set the size of the population to be modeled. For this slider, use a range of 0 through 5000 with increments of 100 and a preset value of 500. Use people as the units label.
- infection-chance is used to set the chance, as a percentage, that a susceptible individual will get infected through an interaction with an infected individual. For this slider, use a range of 0 through 100 with increments of 1 and a preset value of 60. Use percent as the units label.
- init-pct-infected is used to set the initial percentage of the population that is infected. For this slider, use a range of 0 through 100 with increments of 1 and a preset value of 5. Use percent as the units label.

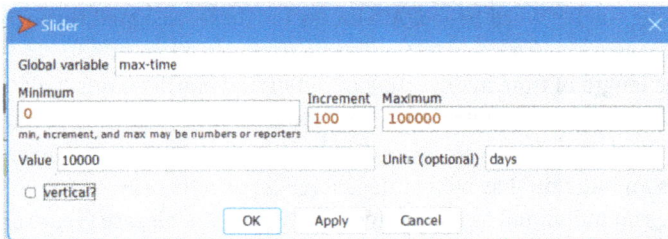

Figure 3.4: Slider for cutoff time in days. The remaining sliders are inserted similarly.

- recovery-chance is used to set the chance, as a percentage, that an infected individual will recover from the disease after a prescribed recovery time in days. For this slider, use a range of 0 through 100 with increments of 1 and a preset value of 60. Use percent as the units label.
- ave-recovery-time is used to set the overall average time in days that it takes for an infected individual to recover from the disease. For this slider, use a range of 0 through 10 with increments of 1 and a preset value of 5. Use days as the units label.

To insert each slider in the Interface window, select Slider from the drop-down menu shown in Figure 3.1 and then click on a blank (white) part of the interface window. Then fill in the necessary details from above as shown in Figure 3.4.

To move a slider to another location (or resize it), right-click on the slider and choose Select and then drag the slider to the desired position (or resize as desired). A slider can also be edited or deleted by right-clicking on it and choosing Edit or Delete. Once this is done, thoughts can be directed to preparing code. But, first save the file (call it, e. g., SIR.nlogo) and continue to do periodic saves along the way.

The first step in preparing code (see Figure 3.5) that will model the behavior of individuals in the community and disease spread is to place variables and/or parameters in certain classes that are referred to in NetLogo syntax as:
- globals, which are variables or parameters that are globally accessible to all procedures in the code;
- breed, also called turtles, which are variable quantities representing agents (or individuals); and
- turtles-own, which are variable quantities associated with attributes of turtles (or agents).

The variables representing global quantities, defined as globals, are
- nb-susceptible, the number of susceptible individuals at a particular time step;
- nb-infected, the number of infected individuals at a particular time step; and
- nb-recovered, the number of recovered individuals at a particular time step.

The variables individuals and individual (which, in NetLogo terminology, are called turtles) are defined using the keyword breed. Attributes for these, assigned through turtles-own, are
- infection-time, the length of time in days that an individual has been infected;
- recovery-time, the number of days after which an individual will recover from the disease;
- infected? identifies an individual as being infected (true) or not (false);
- recovered? identifies an individual as having recovered from the disease (true) or not (false); and
- susceptible? identifies an individual as being susceptible (true) or not (false).

```
;-----------------------------------------------------------------------------
; Define variables, variable attributes, and parameters
;.............................................................................
;
; Quantities entered through sliders
;
; max-time - Cutoff time in days
; pop-size - Total population of community
; infection-chance - Chance of infection as a percentage
; init-pct-infected - Initial percent of infected  individuals
; recovery-chance - Chance of recovery as a percentage
; ave-recovery-time - Average recovery time in days
;
; Global variable quantities
;
; nb-susceptible - Number of susceptible individuals
; nb-infected - Number of infected individuals
; nb-recovered - Number of recovered individuals
;
globals [nb-susceptible nb-infected nb-recovered]
;
; Members of the community, plural and singular (both needed)
;
breed [individuals individual]
;
; Attributes of individuals
;
; infection-time - Time  infected in days
; recovery-time - Recovery time in days
; infected? - Is individual infected?
; recovered? - Has individual recovered?
; susceptible? - Is individual susceptible?
;
turtles-own [infection-time recovery-time infected? recovered? susceptible?]
;.............................................................................
```

Figure 3.5: Code to define the globals, turtles through breed (notice that both plural and singular are needed) and turtles-own, which are used to store attributes assigned to individuals (the turtles).

The next steps involve initializing the population and the canvas for a simulation. As the flowchart in Figure 3.3 suggests, this involves first creating a population of a chosen size, and then assigning each individual in the population certain attributes.

To make things more realistic, attributes are randomly picked and assigned. This means, randomly scatter the population around on the canvas, and then randomly select a chosen percentage of the population to be infected and leave the rest susceptible. The instructions to perform all of these tasks are arranged in three procedures (see Figures 3.6 and 3.7):

- appearance assigns attributes to each individual that determine how they are plotted on the canvas;
- get-people initializes the population and calls appearance to prepare the appearance of individuals on the canvas; and
- Setup calls get-people and gets things ready for a simulation.

```
;------------------------------------------------------------------------------
; Define the canvas setup (initialization) procedures
;..............................................................................
; Procedure to assign plotting attributes for each individual
to appearance
  if susceptible?
     [ set color blue
       set shape "circle"
       set size 0.5]
  if infected?
     [ set color red
       set shape "star"
       set size 1]
  if recovered?
     [ set color green
       set shape "circle"
       set size 0.5]
end
;..............................................................................
; Procedure to create community individuals with randomly assigned initial attributes
to get-people
  ; Create individuals for a population of a given size
  create-individuals pop-size
  [ ; For each individual
    setxy random-xcor random-ycor      ; Place in a randomly selected location
    set susceptible? true              ; Everyone is susceptible to infection
    set infected? false                ; No one is infected
    set recovered? false               ; No one has recovered from infection
    set shape "circle"                 ; Plot individual as a circle
    set color blue                     ; Color plotted point blue
    set size 0.5                       ; Give plotted point a size of 0.5
    ;
    ; Assign a randomly selected and normally distributed recovery time for each individual
    set recovery-time random-normal ave-recovery-time ave-recovery-time / 4
    ;
    ; Make sure the assigned recovery time is not too big
    if (recovery-time > ave-recovery-time * 2) [set recovery-time ave-recovery-time * 2]
    ; and make sure the assigned recovery time is not negative
    if (recovery-time < 0) [set recovery-time 0]
    ;
    ; Infect a randomly selected number of individuals and time infected
    if (random-float 100 < init-pct-infected)
    [ set infected? true
      set susceptible? false
      set recovered? false
      set infection-time random recovery-time]
    ;
    ; Make appropriate changes to appearance attributes
    appearance]
  ;
  ; Initialize counts
  set nb-susceptible count individuals with [susceptible?]
  set nb-infected count individuals with [infected?]
  set nb-recovered 0
end
;..............................................................................
```

Figure 3.6: Code for the appearance and get-people procedures. Notice that spaces need to be placed before and after arithmetic operators. For example, use 2 * x as opposed to 2*x.

There are some NetLogo keywords in Figures 3.6 and 3.7 that are colored. It is worth elaborating on these a bit. Beginning with the blue keywords:

```
; . . . . . . . . . . . . . . . . . . . . . . . . . . . . . . . . . . . . . . . . . . . . . . . . . . . . . . . . . . . . . . . . . . . . .
to Setup
  ; Clear all existing settings
  clear-all
  ;
  ; Set background color to white
  ask patches [set pcolor white]
  ;
  ; Populate the community with individuals having randomly assigned attributes
  get-people
  ;
  ; Reset ticks (aka time-steps) to zero
  reset-ticks
end
; . . . . . . . . . . . . . . . . . . . . . . . . . . . . . . . . . . . . . . . . . . . . . . . . . . . . . . . . . . . . . . . . . . . . .
```

Figure 3.7: Code for the Setup procedure.

- if begins a conditional statement of the form "**if** ⟨*Boolean value is True*⟩ **then** ⟨*perform this task*⟩."
- set represents an assignment statement of the form "**assign** ⟨*a variable*⟩ ⟨*this value*⟩."
- setxy represents an assignment statement of the form "**assign** ⟨*rectangular coordinates*⟩."
- ask represents an instruction statement of the form "**ask** ⟨*patches*⟩ **to do** ⟨*this*⟩."

It will be noticed that values (numeric or otherwise) appear in brown font, and the NetLogo built-in attribute variables to which these values are assigned appear in purple font.

While the purple colored keywords named color, shape, size, and count are self-explanatory, an explanation of the keywords beginning with random is helpful.

- random-xcor and random-ycor are used to generate random values for the *x*- and *y*-coordinates of the implied turtle's (breed's) position.
- Code of the form "random-normal ⟨*arithmetic mean*⟩ ⟨*standard deviation*⟩" extracts randomly generated values from a Normal distribution with given parameters values.
- Code of the form "random-float ⟨*number*⟩" generates random numbers between 0 and the number provided.
- with is used to isolate individuals (turtles) having a certain attribute, such as being susceptible or infected.

Running the Setup procedure (see Figure 3.7) performs the initialization process that sets the stage for a simulation. To facilitate the running of this procedure, place a button for it in the interface window. See Figure 3.8 for directions on how this is done. The final stage is to prepare code to actually perform a simulation of the disease spread.

As suggested by the flowchart in Figure 3.3, procedures for four specific tasks and one encompassing procedure can be constructed to facilitate this; see Figures 3.9 and 3.10:

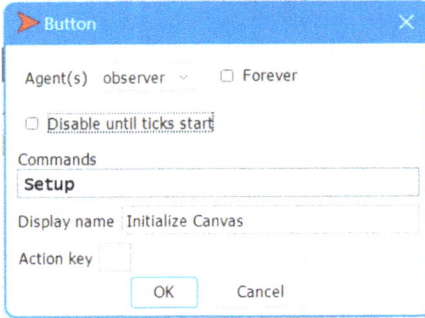

Figure 3.8: Button to initialize the canvas. Just as for sliders, to insert a button in the interface window select Button from the drop-down menu (see Figure 3.1), click on a blank part of the interface window, and then fill in the necessary details. Just as for sliders, buttons can be edited, resized, or moved around.

```
;------------------------------------------------------------------------------
; Define various procedures needed for a simulation
;..............................................................................
; Procedure to move an individual one step in a randomly assigned direction
to move
  rt random-float 360
  fd 1
end
;..............................................................................
; Procedure to decide whether or not an individual gets infected
to infect
  let nearby-uninfected (turtles-on neighbors)          ; Focus on individuals' neighbors
  with [not infected? and not recovered?]
  ;
  ; If at least one neighbor is uninfected, decide whether to infect or not
  ; then assign relevant attributes and set time infected to zero
  if nearby-uninfected != nobody
  [ask nearby-uninfected
    [if random-float 100 >= infection-chance
      [ set infected? true
        set susceptible? false
        set infection-time 0]
    ]
  ]
end
;..............................................................................
; Procedure to decide whether or not an infected individual recovers
to recover
  ; First update infection time
  set infection-time infection-time + 1
  ;
  ; Then check if individual should recover and make appropriate attribute changes
  if (infection-time > recovery-time) and (random-float 100 >= recovery-chance)
    [ set infected? false
      set recovered? true]
end
;..............................................................................
; Update counts of the number of susceptible, infected, and recovered individuals
to update-counts
  set nb-susceptible count individuals with [susceptible?]
  set nb-infected count individuals with [infected?]
  set nb-recovered count individuals with [recovered?]
end
;..............................................................................
```

Figure 3.9: Code containing procedures to be used in the model simulation process.

- move provides instructions on how and to where an individual (turtle) should move on the canvas at each time step.
- infect determines which and when susceptible individuals will get infected.
- recover determines which and when infected individuals will recover.
- update-counts obtains the number of susceptible, infected, and recovered individuals there are at each time step.

Finally, the Go procedure, see Figure 3.10, calls each of the previous four procedures to perform one iteration of a simulation. To wrap things up, just as was done for the initialization, a button to run the simulation should be inserted in the interface window; see Figure 3.11.

```
;--------------------------------------------------------------------------------
; Define the Go procedure to run a simulation
;................................................................................
to Go
   ; Stop simulation if there are no more infections or if it has gone on too long
   if (all? individuals [not infected?]) or (ticks > max-time) [stop]
   ;
   ; For each time step
   ask individuals [ move]                       ; Move individuals
   ask individuals with [infected?]
      [ infect                                   ; Infect candidate susceptibles
        recover]                                 ; Recover candidate infecteds
   ask individuals [appearance]                  ; Update appearance of points
   ask individuals [update-counts]               ; Update counts
   tick                                          ; Advance one tick (day)
end
;................................................................................
```

Figure 3.10: Code to perform one iteration of a simulation.

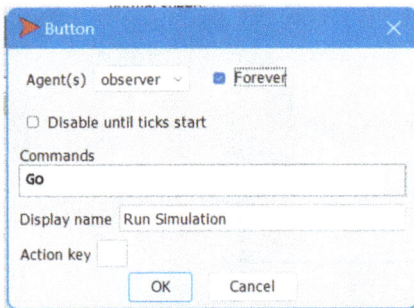

Figure 3.11: Button to run a simulation after initializing the canvas. Notice that a checkmark appears by Forever. This instructs NetLogo to keep performing iterations of the simulation. The exit conditions in the Go procedure ensures that the iterations will stop at some point. A simulation can also be stopped at any time by simply clicking on this button again.

Two blue keywords in the Go procedure (see Figure 3.10) need some explanation:
- Code of the form "ask ⟨*turtle name*⟩ [*task(s)*]," in a sense, "asks" all turtles having the specified name to undergo the task(s) specified. Analogous code is used on patches in the Setup procedure.
- tick advances the simulation by one time step.

3.5.3 Monitors and plots

For purposes of this example, it would be informative to include two forms of simulation output features in the interface window:
- Insert monitors to display counters for time steps (see Figure 3.12) and the numbers of individuals in the susceptible, infected and recovered pools at each time step—use nb-susceptible, nb-infected, and nb-recovered, respectively, as reporters; and
- Insert a time-series plot of the percentage of individuals in each of the three categories; see Figure 3.13.

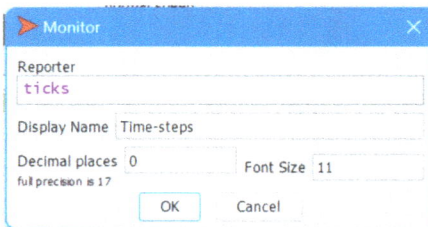

Figure 3.12: To insert a monitor in the interface window select Monitor from the drop-down menu, see Figure 3.1, and then click on a blank part of the interface window. Then fill in the necessary details as shown. The process to edit, resize, or move a monitor around is the same as for sliders.

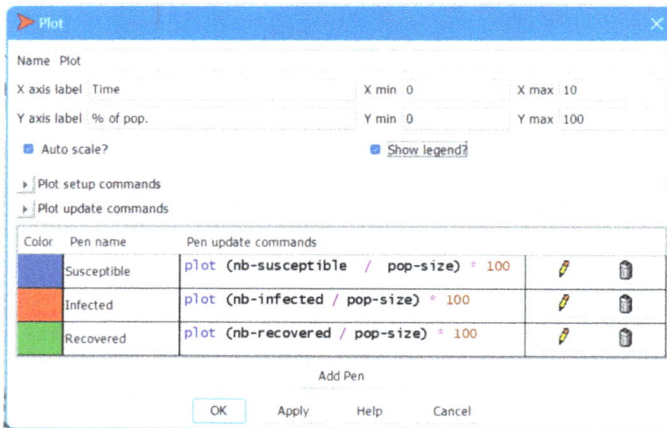

Figure 3.13: To insert a plot in the interface window select Plot from the drop-down menu, see Figure 3.1, and then click on a blank part of the interface window. Then fill in the necessary details as shown above. Checking Auto scale? ensures that the plots produced are complete, and checking Show legend? results in a legend for the three plots produced. Plot colors can be customized by clicking on each of the colors, pen names can be edited, and pen update commands are entered as NetLogo code. As before, the process to edit, resize, or move a plot around is the same as for sliders.

Assuming all goes perfectly and no coding errors or other issues occur, the reorganized interface window might have the appearance of Figure 3.14. Be sure to save the file before continuing.

3.5.4 Simulations and observations

First, click on the Initialize Canvas button, Figure 3.15 shows one possible initial setup. If another initialization is performed, the number of infected individuals (up to 1% of the population) and the placement of all individuals will change. It may also be that the random process used will not infect any individuals! Clicking on the Run Simulation button will start a simulation, and simulations may be stopped at any time by clicking on this button again. Figure 3.16 shows one possible outcome for the initialized setup in Figure 3.15, compare the plot produced here with that in Figure 2.17. Another simulation can be performed by clicking on the Initialize Canvas then Run Simulation buttons again.

Out of curiosity, change the view updates (see Figure 3.1) choice to on ticks and perform a few simulations. Notice that when this is done the process slows down and changes in the canvas can be followed more easily. Adjusting the speed from normal speed to slower makes it even easier to follow changes.

If there is an interest in looking at one iteration of the simulation at a time, right-click the Run Simulation button, choose Edit..., then uncheck the Forever box, then click on OK. Each click on the Run Simulation button then results in exactly one iteration (time step) being performed, and changes to the canvas can be followed exactly.

3.5.5 Exploratory exercises

Once an agent-based model has been prepared in NetLogo, it becomes possible to have fun with it by exploring a wide range of "what if" scenarios and hypothesizing connections between the outcomes and the environment within which the disease spreads. Here are some possible explorations that may be conducted—do not save any changes made to the model while going through these exercises:

1. With the original parameter (slider) settings, and using the Forever setting in the simulation button with a preferred simulation speed and view updates choice (see Figure 3.1), perform a few simulations—pay close attention to the starting number of infected individuals, placement and movement of the infected individuals, and the appearance of the curves in the plots produced. Repeat simulations as needed for each of the first three questions:

    (a) On average, where does the horizontal position of the peak of the infected curve appear to fall in relation to the intersection point of the susceptible and recovered curves?

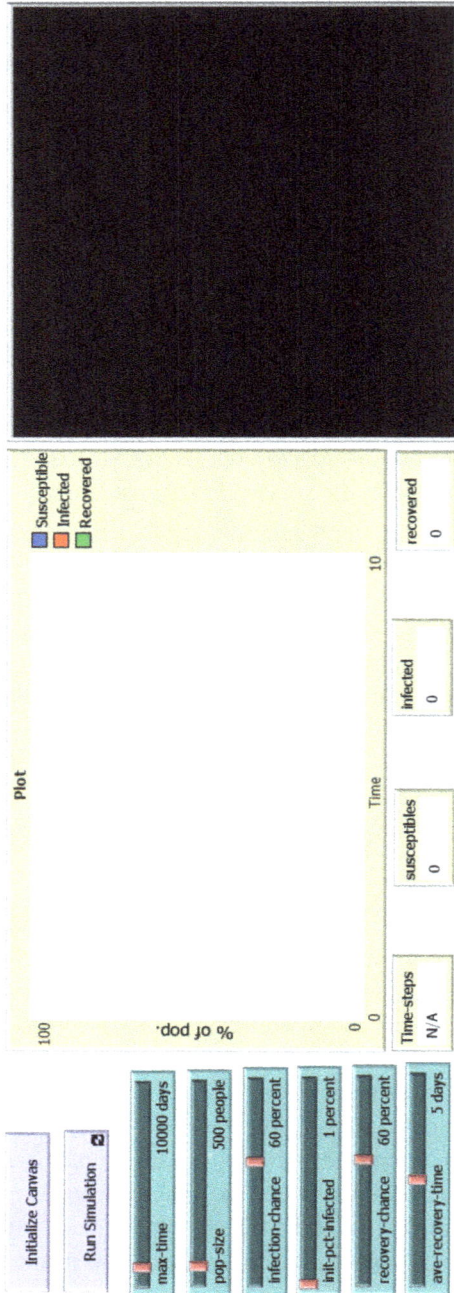

Figure 3.14: Completed interface window. Note that the canvas and the various inserts have been resized and rearranged.

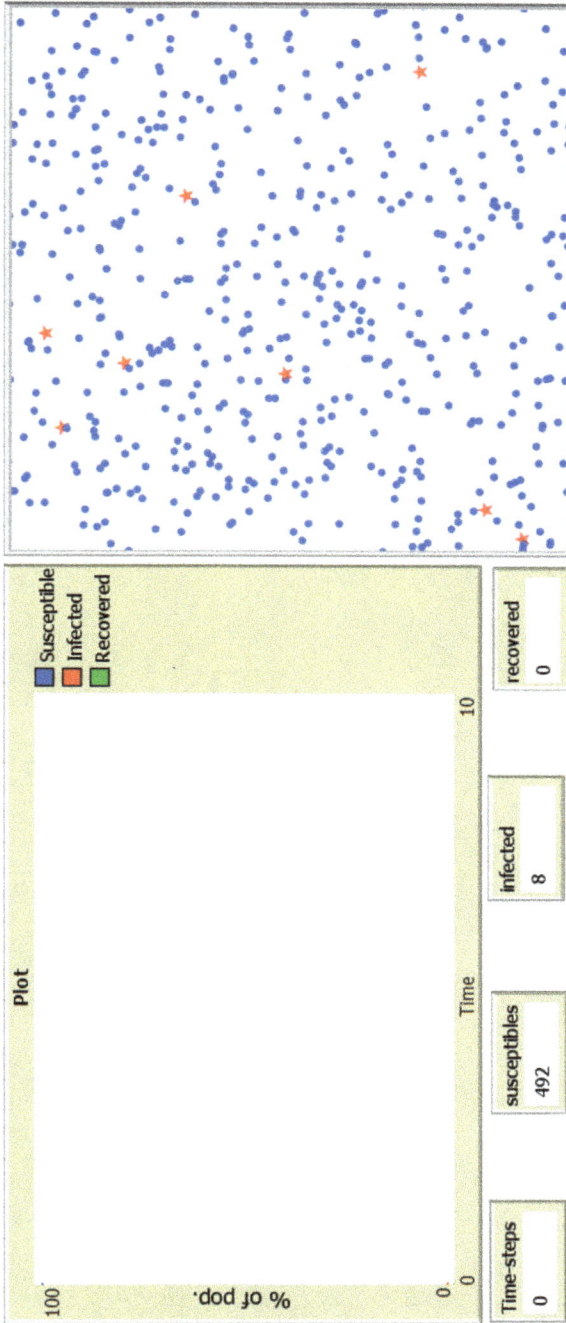

Figure 3.15: An initialized canvas. Red stars represent infected individuals and blue dots represent susceptible individuals.

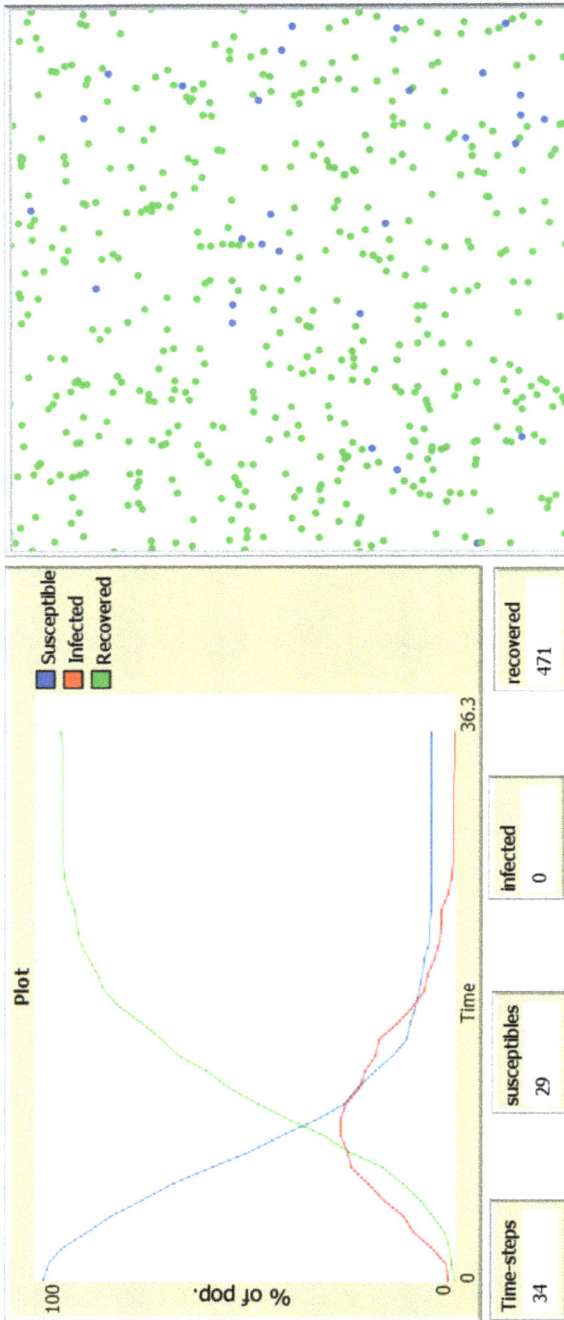

Figure 3.16: A completed simulation.

(b) Does the size of the (randomly assigned) initial number of infected individuals seem to influence observations for the previous question?

(c) How does the placement of infected individuals among the rest of the population seem to influence simulation outcomes?

(d) What do observations/answers for the previous questions seem to suggest about the spread of this disease in a "real-world" scenario?

2. Keeping all *except* the `init-pct-infected` sliders the same (the original setting), perform simulations for *increasing* percentages of initial infections—try slider positions of 10 %, 15 %, 20 %, ...

(a) On average, where does the horizontal position of the peak of the infected curve appear to fall in relation to the intersection point of the susceptible and recovered curves?

(b) Does an increasing percentage of initial infections appear to strengthen or weaken any observations and conclusions made from the previous set of exercises? How?

(c) Are there any noticeable patterns in the total number of time steps as the initial percentage of infected individuals is increased?

(d) What does all of this seem to suggest for a "real-world" scenario?

3. Set the initial number of infected individuals at 5 %. Then, with all the remaining sliders, *except* the `infection-chance` slider, in the original setting perform simulations for lower and higher chances of infection—try slider positions of 45 %, 50 %, 55 %, 60 %, 65 %, 70 %, and 75 %.

(a) Are there any noticeable patterns that appear in the plotted curves as the chance of infection increases (e. g., from 45 % to 75 %)?

(b) Are there noticeable patterns in the total number of time steps as the chance of infection increases (e. g., from 45 % to 75 %)?

(c) What does all of this seem to suggest for a "real-world" scenario?

(d) What might be "real-world" reasons for decreased or increased chances of infection?

4. Set the initial number of infected individuals at 5 %. Then, with all the remaining sliders, *except* the `recovery-chance` slider, in the original setting perform simulations for lower and higher chances of recovery—try slider positions of 45 %, 50 %, 55 %, 60 %, 65 %, 70 %, and 75 %.

(a) Are there any noticeable patterns that appear in the plotted curves as the chance of recovery increases (e. g., from 45 % to 75 %)?

(b) Are there noticeable patterns in the total number of time steps as the chance of recovery increases (e. g., from 45 % to 75 %)?

(c) What does all of this seem to suggest for a "real-world" scenario?

(d) What might be "real-world" reasons for decreased or increased chances of recovery?

5. What happens if the chance of infection is decreased, and if the chance of infection is increased?

6. Set the initial number of infected individuals at 5 %. Then, with all the remaining sliders, *except* the `ave-recovery-time` slider, in the original setting perform simulations for lower and higher average recovery-times—try slider positions of 2, 3, 4, 5, 6, 7, and 8 days.
   (a) Are there any noticeable patterns that appear in the plotted curves as the average recovery time increases (e. g., from 2 to 8 days)?
   (b) Are there noticeable patterns in the total number of time steps as the average recovery time increases (e. g., from 2 to 8 days)?
   (c) What does all of this seem to suggest for a "real-world" scenario?
   (d) What might be "real-world" reasons for decreased or increased average recovery times?
7. What scenario appears to maximize the total percentage of the population that gets infected (hence recovers)?
8. What scenario appears to minimize the total percentage of the population that gets infected (hence recovers)?
9. Does increasing the size of the population affect the dynamics of the disease-spread in any way? How?
10. It turns out that NetLogo can get "unhappy" about what it is asked to do. Try increasing the population size (by 100s); see what eventually happens.
    (a) What does this suggest is worth keeping in mind when using NetLogo for agent-based modeling?
    (b) What might be one way to accommodate larger population sizes? Hint: Try making changes to the *World* settings. What are changes that appear to help the cause?
11. Recall the "world wrap" feature mentioned in Section 3.4.2 (see Figure 3.2). Reset all parameter to their original values (see Figure 3.14), change the `view updates` to `on ticks`, use the simulation speed slider to slow things down a bit, and then run a simulation and make a mental note of what happens when individuals encounter an edge of the canvas. Then open the `Settings` pop-up and uncheck the two *world wrap* boxes and run some more simulations. What happens when individuals encounter the edges of the canvas now? Does this suggest anything useful for the model settings?

Before continuing, close the `SIR.nlogo` file *without* saving changes made while going through the above exercises, then reopen it and create a copy by saving it under a different name (e. g., as `SIR2.nlogo`).

The code for *epiDEM* example, [125], contains a procedure to estimate the reproductive number, $\mathcal{R}_0$, for a modeled disease. Here is a coding challenge exercise:

12. Adapt the code in the (new) `SIR2.nlogo` file/model to accommodate the procedure in the *epiDEM* model code that computes an estimate of $\mathcal{R}_0$.

(a) In the process of adapting the code, it is helpful to make a note of issues that arise, for future reference. What are the main ones?

(b) Explore if and how changes to parameter settings affect the estimated value of $\mathcal{R}_0$.

3.5.6 Extensions and other ideas

As was the case for the SIR compartmental model, based on the assumption that no births, deaths, emigration, or immigration occur, this model is clearly best suited for a closed community over a relatively short time period. This, combined with the assumption that reinfections do not occur due to acquired immunity, implies that the agent-based SIR model also always predicts the eventual decline of a disease it is used to model. As mentioned before, this may not always be realistic, and for this reason other models might be preferred.

The strategy used for the SIR agent-based model can be used to create alternative agent-based models for infectious diseases by extending the SIR model through asking, and answering directed questions. For example:
- Which characteristics/attributes serve as useful means of partitioning the population? Notice the similarity to compartmentalizing in the previous compartmental models.
- What are expected roles that these attributes would play in modeling the dynamics of the disease?
- What effects will interactions between individuals with different attributes have?
- Are there any parameters that will serve as important global quantities?
- What might the expected dynamics of the system being modeled look like? Here is where constructing a flowchart (or a collection of subflowcharts) will help.
- What might be useful/informative plots and monitors?

Once a model has been designed, other questions can be asked. For example:
- For which parameters/globals should sliders be included? Why?
- What useful information would varying each parameter provide?
- What would a meaningful goal of performing an analysis of the model through parameter variations be?

Agent-based versions of the previously listed infectious disease models (see Section 2.4.7) can be constructed by extending the code used for the SIR model. In each case at the least one or more procedures that control the movement of individuals from one class (via attributes) to the next would need to be incorporated into the code.
- **SIRS** (Susceptible-Infected-Recovered-Susceptible): This is similar to the SIR model, except in that a recovered individual is assumed to gain short-term immunity and

eventually moves back into the susceptible pool. The manner in which this movement will occur can be controlled by an additional procedure.

- **SIRV** (Susceptible-Infected-Recovered-Vaccinated): This too is identical to the SIR model in all respects except in that members from the susceptible pool who gain at least short-term immunity through vaccination are accounted for. Individuals who have recovered are assumed to have gained at least short-term immunity. This suggests the need for a fourth attribute, vaccinated?, introduced through turtles-own and a procedure to manage the new class of individuals.

- **SIRD** (Susceptible-Infected-Recovered-Deceased): This is identical to the SIR model in all respects except in that infected individuals may recover and gain immunity, or they may become deceased. This suggests the need for the corresponding individual to die and be removed from the population pool. So, a new procedure to manage this will be needed.

- **SIS** (Susceptible-Infected-Susceptible): This model assumes the absence of immunity (even after recovering from an infection), and that an infected individual who recovers moves back into the susceptible pool. Here, one less turtle attribute, recovered?, is needed and the recover procedure (see Figures 3.9 and 3.10) can be adapted to manage this.

- **SEIS** (Susceptible-Exposed-Infected-Susceptible): This is similar to the SIS model, except in that there is an in-between period of time when a susceptible individual has been exposed to the disease but does not display visible signs of infection. During this period of time, even though the exposed individuals do not display visible signs of infection, they are assumed to be infectious. This suggests the need for an attribute exposed? through turtles-own, and the removal of the attribute recovered? and procedure recover. A procedure to manage exposed individuals would need to be included.

- **SIRVD** (Susceptible-Infected-Recovered-Vaccinated-Deceased): This model combines elements of the SIR, SIRV, and SIRD models. This suggests the need for the attribute vaccinated? through turtles-own and the ability of an individual to die. Procedures to manage these would need to be included in the code.

- **MSIR** (Maternal-Susceptible-Infected-Recovered): Here, the term "maternal" identifies individuals (typically babies) who have gained passive immunity from maternal antibodies. So, this model accounts for another form of immunity. This suggests the need for a fourth attribute, maternal?, through turtles-own and a procedure to manage this would be needed. Notice that this suggests a need to accommodate births.

- **SEIR** (Susceptible-Exposed-Infected-Recovered): This too is similar to the SIR model, except in that there is an in-between period of time when a susceptible individual has been exposed to the disease but does not clearly display visible signs of infection. During this period of time, even though the exposed individuals do not display visible signs of infection, they are assumed infectious. This suggests the need for an attribute exposed? through turtles-own and a procedure to manage it.

- **MSEIR** (Maternal-Susceptible-Exposed-Infected-Recovered): This is similar to the SEIR model, except in that there is a proportion of the population that has acquired immunity through maternal antibodies. This suggests the need for attributes exposed? and maternal? through turtles-own and appropriate procedures to manage these.
- **MSEIRS** (Maternal-Susceptible-Exposed-Infected-Susceptible): This follows the MSEIR model, except in that recovered individuals become susceptible again after a period of time. Relevant attributes and procedures to manage them will need to be included in the code.

Again, one could ask further questions. For example:
- Can births and deaths due to natural causes (those unrelated to infection) be brought into the picture?
- Would it be meaningful to bring in births and deaths due to natural causes into the picture?
- Does age play a role in infection rates? Would this be worth exploring?
- Does age play a role in recovery rates? Would this be worth exploring?
- Does age play a role in deaths due to infection? Would this be worth exploring?
- How might quarantining be brought into the picture? Would this be worth exploring?
- Can immigration and emigration be brought into the picture? Would this be worth exploring?
- How could restricting the movement of individuals in a community to streets, and to specific locations such as schools, parks, markets, offices, and homes be accomplished? Would this be worth exploring?
- It is possible that natural spatial (or cultural) boundary exist in the community that discourage or restrict (but not necessarily block) the movement of individuals from one part of the community to another. How might this be addressed? Would this be worth exploring?
- It is possible that individuals in the community experience periods of temporary isolation, as in periods of sleep. How might this be introduced? Would this be worth exploring?

As always it is helpful to remember that a good understanding of the underlying biological or epidemiological background information will play a big part in helping achieve a reasonably realistic simulation.

3.6 Coding tips

As anyone who has engaged in any type of a coding/programming exercise will attest to, periodic encounters with (often very annoying) glitches should be expected. View these as interesting challenges rather frustrating hurdles.

For beginners, the best way to set the stage for constructing an agent-based model (hence, prepare NetLogo code) is to make good use of the ODD protocol to establish a clear and complete understanding of the problem in question, and the processes that will be involved. This, combined with a carefully prepared and clear flowchart, helps a lot in figuring out the big picture as well as meaningful points at which self-contained modules will be useful. This being said, there are some coding tips that can be offered.

- An efficient and effective way to get started with coding in NetLogo is to take advantage of code in freely available and functioning NetLogo models that can be adapted to a different, but equivalent, story.
- A keyword search of the form, for example, "turtles" in the *Beginner's Interactive NetLogo Dictionary* at

    https://ccl.northwestern.edu/netlogo/bind/

    is a quick way to go directly to the syntax definition and possible examples of its use.
- A Google search of the form, for example, "turtles-own in NetLogo" leads directly to the relevant location in the *NetLogo Dictionary* at

    https://ccl.northwestern.edu/netlogo/docs/index2.html

- Practice modular coding as much as is possible. This means, identify specific tasks that need performing and that can be represented by an appropriately named procedure. Using a modular approach in coding with plenty of procedures for specific tasks rather than one long program makes tracking down logical errors more efficient.
- It turns out that NetLogo has the ability to spot errors and/or inconsistencies in code syntax, and alerts to issues will very often pop up as soon as they occur. Watch out for these alerts. Fixing inconsistencies along the way helps save time down the road.
- The Check button in the Code window is very helpful in catching syntax errors (a red checkmark or warning/error message indicates problems). Use this each time a procedure is completed. If there are any issues in the procedure, an informative error message pops up to help in fixing the issue. Click on the Check button again once changes to the code are made.
- While not necessary, it is helpful to arrange procedures in the order that their needs appear in the flowchart for an agent-based model. Notice that in the previous example the Setup and Go procedures call preparatory procedures in a manner that at least closely follows the broad tasks identified in the flowchart.

Logical errors in code often do not prevent a program from running. However, code with logical errors will produce misleading or nonsense results. The Check button will not help in detecting logical errors, so tracking down such errors is the responsibility of the coder/modeler. To avoid having to track down large numbers of logical errors, it is best to keep the expected "flow" of the program in mind while checking the soundness of each procedure along the way—refer to the model's flowchart and focus on describing the overall flow through a collection of main steps (or procedures). That is, how do the outcomes of individual procedures contribute to other procedures, and to the whole picture?

3.7 A predator–prey model

Consider looking at a slightly enhanced version of the predator–prey scenario described in Section 2.5 from an agent-based modeling point of view. Here also, consider tracking the population sizes of a single prey species and its sole predator species in a closed ecosystem, but create limitations on the prey food-source. The strategy here follows that used in developing the previous SIR agent-based model. Again, keep the ODD protocol while going through the following.

3.7.1 Setting the stage

First, establish some assumptions for the system—recall the assumptions for the corresponding compartmental model in Section 2.5, paying close attention to the similarities and differences between the two. The goal is to track the predator and prey population sizes over some time period, $0 \leq t \leq T$ time steps:

- Instantaneous and uniform mixing of the two species does not occur. However, both species do move around freely in their shared ecosystem, but they lose energy in the process.
- The reproductive cycles of the predator and the prey species are the same.
- Predator and prey interactions may or may not occur. For every interaction between a predator and a prey that does occur, there is a chance that the predator will catch (and eat) the prey.
- If a predator succeeds in catching a prey, it gains energy.
- A predator has some chance of reproducing (depending on the female–male ratio).
- If a predator's energy-level becomes zero, it dies.
- The two ways in which a member of the prey population can die is if it is caught by a predator, or if its energy-level becomes zero due to starvation.
- A member of the prey population increases its energy-level by eating food, of which there is only one type.
- A member of the prey population has some chance of reproducing (again, depending on the female–male ratio).
- The prey food-source in any patch of the ecosystem can be exhausted, but it will replenish after some time period.

The flowchart in Figure 3.18 provides one way in which the process can be summarized at the individual level for the predator, the prey, and the prey's habitat (the patches).

3.7.2 Preparing the canvas and the code

Ideas and code for this example are adapted from the second example in [5]. See also the "Wolf Sheep Predation" sample model provided in "Tutorial #1" of the NetLogo User Manual, [121]. Start by saving the (empty) model file, for example, named as PP.nlogo.

For this example, use the on ticks mode for view updates and make some of adjustments to the Settings. Use maximum x and y coordinates of 48 with a patch size of 5 and a frame rate of 20; see Figure 3.17.

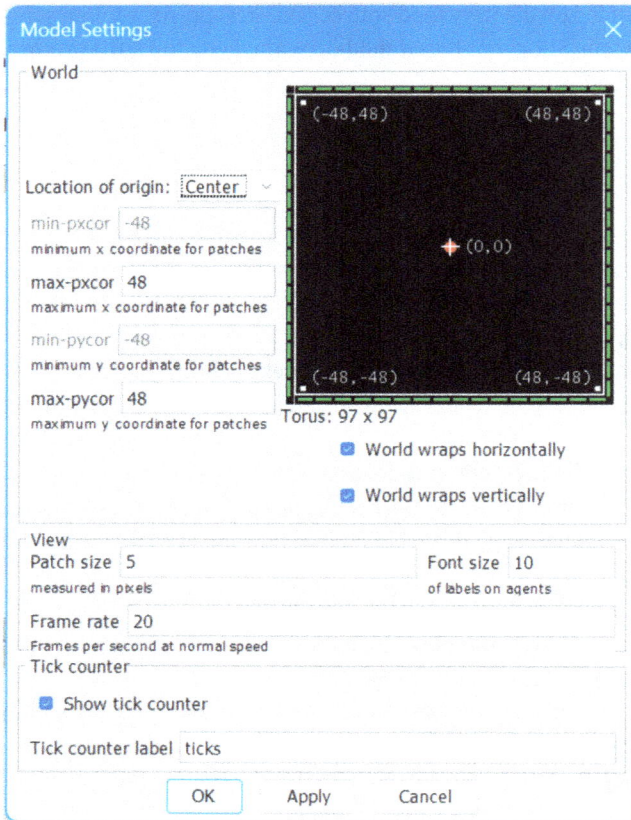

Figure 3.17: World settings for the predator–prey model.

Keeping the listed assumptions for the scenario in question and Figure 3.18 flow-chart in mind, a good place to start would be to decide on what global values would best be introduced through sliders, inserted in the interface window in the manner described for Figure 3.4. Some obvious candidates would be:

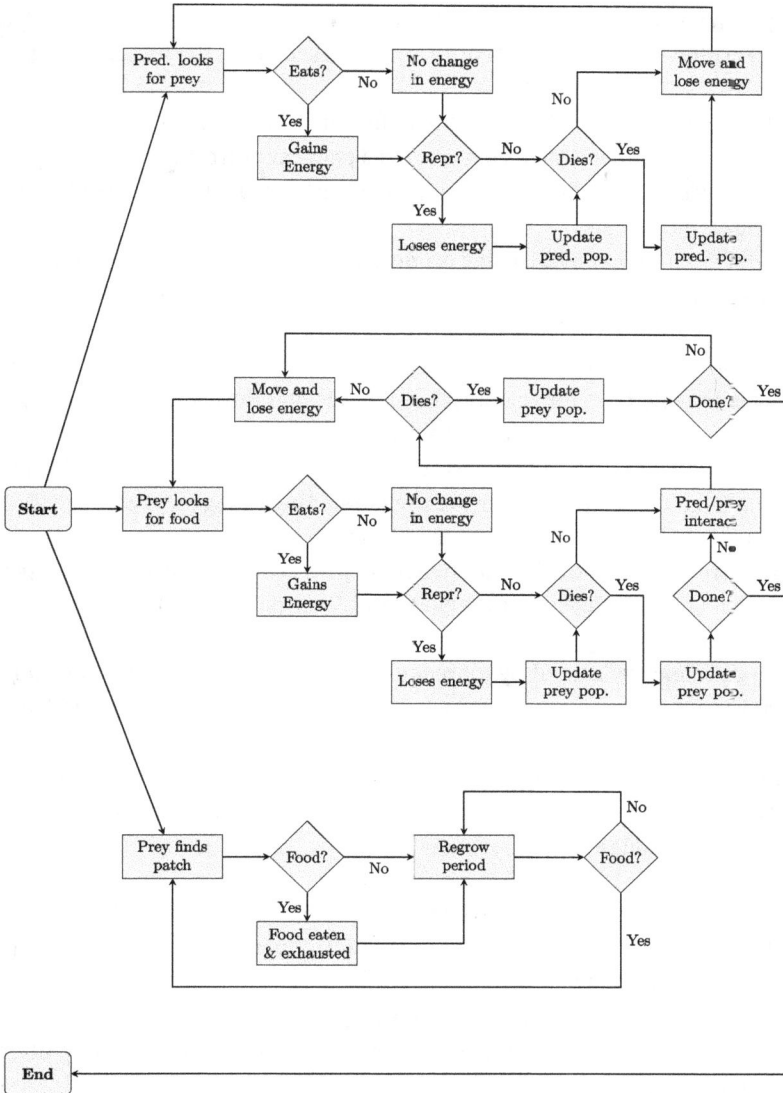

Figure 3.18: A flowchart representing one possible description of the dynamics of a predator–prey system in which the prey food source can be depleted (and replenished). The simulation stops when there are no prey remaining, or if the Run Simulation button is clicked on again.

- max-time, cutoff number of time steps for simulations if predators die off. Use a range of 0 through 10,000 with increments of 1000 and a value of 1000 as the initial setting.
- initial-prey-pop, the initial prey population size. Use a range of 0 through 500 critters with increments of 20 and a value of 100 as the initial setting.

- `initial-predator-pop`, the initial predator population size. Use a range of 0 through 250 critters with increments of 10 and a value of 50 as the initial setting.

Additional sliders that can add to the flexibility of this model could be:
- `prey-brood-size`, number of offspring produced per year by a candidate individual prey. Use a range of 0 through 15 babies with increments of 1 and a value of 1 as the initial setting.
- `pred-brood-size`, number of offspring produced per year by a candidate predator. Use a range of 0 through 10 babies with increments of 1 and a value of 1 as the initial setting.
- `pct-prey-reproducers`, the percentage of prey that will reproduce. This depends on the female to male ratio for the prey. Use a range of 0 through 25 % with increments of 1 and a value of 4 as the initial setting.
- `pct-pred-reproducers`, the percentage of predators that will reproduce. This depends on the female to male ratio for the predators. Use a range of 0 through 25 % with increments of 1 and a value of 5 as the initial setting.
- `prey-food-gain`, energy gained by an individual prey from food in some units per patch consumed. Use a range of 0 through 10 units per patch with increments of 1 and a value of 5 as the initial setting.
- `pred-food-gain`, energy gained by a predator from food in some units of energy per prey consumed. Use a range of 0 through 25 units/prey with increments of 1 and a value of 20 as the initial setting.
- `patch-replenish-time`, the average time it takes for a patch to regrow and replenish itself with prey food. Use a range of 0 through 50 with increments of 5 and a value of 30 as the initial setting.

Initial values used in the sliders are those used in the previously mentioned Wolf-Sheep example. The remaining variables and attributes are introduced through code; see Figure 3.19. Global quantities include:
- `total-time`, the total time at end of each time step.
- `nmbr-prey` and `nmbr-pred`, the number of prey and predators, respectively, after each time step.
- `total-patches`, the total number of patches on the canvas.
- `nmbr-brown-patches`, the number of brown patches at the end of each time step.
- `pct-brown-patches`, the percent of the canvas that contains brown patches at the end of each time step.

The two species (turtles) are introduced through `breed` as:
- `predators` and `predator`, representing the plural and singular, respectively.
- `preys` and `prey`, representing the (intentionally misspelled) plural and singular, respectively.

```
;-------------------------------------------------------------------------------
; Define variables,variable attributes, and parameters
;...............................................................................
; Quantities entered through sliders
;
; max-time - cutoff time
; initial-prey-pop - initial prey population size
; initial-pred-pop - initial predator population size
; prey-brood-size -  prey offspring per critter per time step
; pred-brood-size -  predator offspring per critter per time step
; pct-prey-reproducers - percent prey that will reproduce
; pct-pred-reproducers - percent predators that will reproduce
; prey-food-gain - units of energy gained by prey per patch of food
; pred-food-gain - units of energy gained by predator per prey caught
; patch-replenish-time - time for a depleted patch to regrow prey food
;
; Global variable quantities
;
; total-time - time counter
; nmbr-prey - counter for number of prey
; nmbr-pred - counter for number of predators
; total-patches - total number of patches on canvas
; nmbr-brown-patches - counter for number of brown (depleted) patches
; pct-brown-patches - Percent of patches colored brown
globals[total-time nmbr-prey nmbr-pred total-patches nmbr-brown-patches pct-brown-patches]
;
; Species, as plural and singular (one for each)
;
breed[predators predator]
breed[preys prey]               ; Intentionally misspelled
;
; Breed and patch attributes
;
turtles-own[energy-left]
patches-own[regrow-time-left];
;-------------------------------------------------------------------------------
```

Figure 3.19: Code defining globals, turtles (as breeds), turtle and patch attributes.

The species attribute energy-left (species energy remaining at each time step) is introduced through turtles-own, and regrow-time-left (time remaining for a patch to replenish itself) is introduced through patches-own.

The next task is to prepare procedures needed to initialize the canvas; see Figure 3.20. These include:

- initialize-patches randomly colors patches green or brown, then assigns a random remaining replenishment time to brown patches.
- initialize-prey generates the initial prey population of the provided size, provides plotting characteristics, assigns a random amount of remaining energy to each individual, and then places them at randomly selected points on the canvas.
- initialize-predators generates the initial predator population of the provided size, provides plotting characteristics, assigns a random amount of remaining energy to each individual, and then places them at randomly selected points on the canvas.

```
;--------------------------------------------------------------------------------
; Define procedures to initialize the canvas
;................................................................................
;
; Initialize the patches
to initialize-patches
  ask patches
  [ set pcolor one-of [green brown]
    ifelse (pcolor = brown) [set regrow-time-left random patch-replenish-time]
                            [set regrow-time-left patch-replenish-time]
  ]
end
;................................................................................
; Initialize the prey population
to initialize-prey
  create-preys initial-prey-pop
  [ set shape "circle"
    set size 1
    set color white
    set energy-left random (2 * prey-food-gain)
    setxy random-xcor random-ycor]
end
;................................................................................
; Initialize the predator population
to initialize-predators
  create-predators initial-pred-pop
  [ set shape "star"
    set size 2
    set color black
    set energy-left random (2 * pred-food-gain)
    setxy random-xcor random-ycor]
end
;................................................................................
; Initialize all counters
to initialize-counters
  set total-time 0
  set nmbr-prey count preys
  set nmbr-pred count predators
  set total-patches count patches
  set nmbr-brown-patches count patches with [pcolor = brown]
  set pct-brown-patches (nmbr-brown-patches / total-patches) * 100
end
;................................................................................
; Combine in the setup procedure to initialize the canvas
to Setup
  clear-all
  initialize-patches
  initialize-prey
  initialize-predators
  initialize-counters
  reset-ticks
end
;--------------------------------------------------------------------------------
```

Figure 3.20: Procedures to perform tasks involved in initializing the canvas. The Setup procedure is called by the Initialize Canvas button, which is placed in the interface window as described in Figure 3.8.

- initialize-counters initializes counters for each of: the time steps completed; the total number of patches on the canvas; the initial number of prey and predators; and the number of brown patches on the canvas. It also computes the initial percentage of brown patches.
- Setup calls the above procedures to initialize the canvas.

As done in Section 3.5.2, an `Initialize Canvas` button to run the `Setup` procedure is placed in the interface window.

Procedures that provide instructions for specific tasks in each iteration of a simulation (to be called in the `Go` procedure) (see Figures 3.21 and 3.22) are:

- `move` instructs individuals from each species to move one step in a randomly selected direction.
- `reproduce-prey` randomly selects prey with enough energy to reproduce a specified number of offspring.
- `reproduce-pred` randomly selects predators with enough energy to reproduce a specified number of offspring.
- `predation` gets each predator to randomly select one nearby individual prey (if any) to be eaten, resulting in a prey death and the predator gaining a specified amount of energy from the meal.
- `death` gets members of each species with zero energy to go away (die).
- `prey-feeds` gets prey to feed on the patch within which it stands when the patch has food. In the process, food in the patch is depleted and the prey gain energy from the food.
- `replenish-patch` replenishes depleted prey food in brown patches over time.
- `update-counters` updates the various global counters.

Before preparing the `Go` function, which performs simulations, one might be interested in including a switch that turns a tracker on (or off) for the movement of the predators in the interface window. To insert a switch to activate a tracker for the movement of predators in the interface window, select `Switch` from the drop-down menu shown in Figure 3.1. Then click on an empty spot in the interface window and enter the global quantity `track-predator?`. This is used in a conditional statement that is included in instructions for the predator in the `Go` procedure; see Figure 3.22. The editing, resizing, and moving of switches is as for other objects contained in the interface window.

At this point, it is useful to draw attention to some NetLogo syntax encountered in the code shown in Figures 3.19–3.22. These include:

- `patches`, `patches-own`, and `pcolor` are associated with patches on the canvas, their attributes, and color.
- `turtles` and `turtles-own` are associated with turtles (through `breed`) and their attributes.
- `lt` and `rt` are used to instruct a turtle to turn a given number of degrees to the left or right, `fd` instructs a turtle to move a given number of steps forward.
- `pen-down` (or `pen-up`) instructs NetLogo to draw lines that show the path followed by a turtle (or not).

There are other terms that have been encountered so far in this and the previous example, such as `random`, `let`, `one-of`, `nobody`, `not`, and `!=`. Look up information about these (or others of interest) in the NetLogo dictionary.

```
;---------------------------------------------------------------
; Define procedures to perform a simulation
;...............................................................
; Move in a random direction
to move
  rt random 50
  lt random 50
  fd 1
end
;...............................................................
; Pick random prey with enough energy  to reproduce
to reproduce-prey
  if (energy-left > energy-left / 2) and (random-float 100 < pct-prey-reproducers)
    [set energy-left (energy-left / 2)            ; Lose energy in reproducing
     hatch prey-brood-size [rt random 360 fd 1]]  ; Produce offspring and move
end
;...............................................................
; Pick random predators with enough energy  to reproduce
to reproduce-predator
  if (energy-left > energy-left / 2) and (random-float 100 < pct-pred-reproducers)
    [set energy-left (energy-left / 2)            ; Lose energy in reproducing
     hatch pred-brood-size [lt random 360 fd 1]]  ; Produce offspring and move
end
;...............................................................
; Pick random prey to be eaten
to predation
  let victim one-of preys-here                ; Look around
  if victim != nobody                         ; If there is a nearby victim
    [ask victim [die]                         ; Eat it
     set energy-left energy-left + pred-food-gain]  ; Gain energy
end
;...............................................................
; If critter has no energy, it dies
to death
  if energy-left <= 0 [die]
end
;...............................................................
; Prey feeds on its food source and gains energy, but depletes food
to prey-feeds
  if pcolor = green
   [set pcolor brown
    set energy-left energy-left + prey-food-gain]
end
;...............................................................
; Depleted prey food source replenishes after some given time
to replenish-patch
  if (pcolor = brown)
   [ifelse (regrow-time-left <= 0)
     [set pcolor green
      set regrow-time-left patch-replenish-time]
     [set regrow-time-left regrow-time-left - 1]  ]
end
;---------------------------------------------------------------
```

Figure 3.21: Procedures to perform tasks involved in a simulation of the predator–prey system.

The final task remaining is to include plots in the interface window; see Figures 3.23 and 3.24. Plots are inserted as described for the SIR model in Section 3.5.3. Once this is done, the starting interface window may have the appearance of Figure 3.25 and a sample simulation may end up as in Figure 3.26.

```
; -----------------------------------------------------------------------------
; Procedure to Update counters after each simulation
; .............................................................................
to update-counters
  set total-time ticks
  set nmbr-prey count preys
  set nmbr-pred count predators
  set nmbr-brown-patches count patches with [pcolor = brown]
  set pct-brown-patches (nmbr-brown-patches / total-patches) * 100
end
; .............................................................................
;
; -----------------------------------------------------------------------------
; Run the simulation
; .............................................................................
to Go
  if (nmbr-prey = 0) or ((nmbr-pred = 0) and (total-time > max-time)) [stop]
  ask preys [
    move
    set energy-left energy-left - 1
    prey-feeds
    death
    reproduce-prey]
  ask predators[
    ifelse track-predators? [pen-down] [pen-up]    ; Follow predator movement, or not
    move
    set energy-left energy-left - 1
    predation
    death
    reproduce-predator]
  ask patches [replenish-patch]
  tick
  update-counters
end
; -----------------------------------------------------------------------------
```

Figure 3.22: Code for each iteration of a simulation. A button to run the Go procedure is inserted in the interface window as described in Figure 3.11.

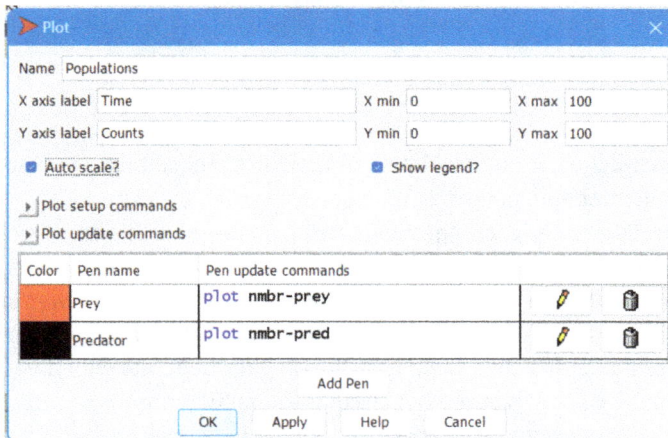

Figure 3.23: Setting for a plot of population counts against time.

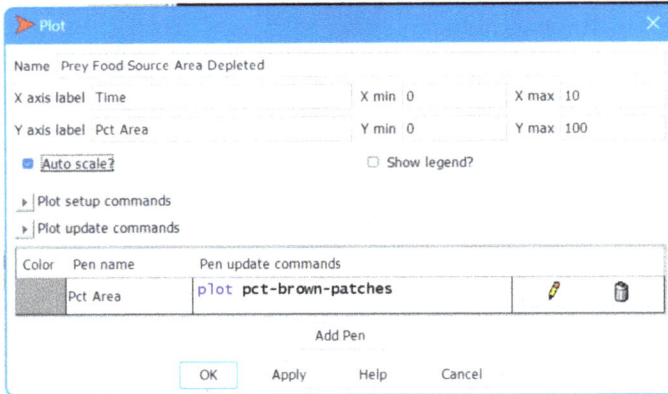

Figure 3.24: Plot of the percentage of brown patches against time.

3.7.3 Simulations and observations

Simulations are performed as for the SIR model. Click on the `Initialize Canvas` button a few times, and notice how the appearance of the patches and the placement of predators (black stars) and prey (white bullets) changes each time. Sometimes they are widely spread out, and sometimes clusters of one or both species are produced. Clicking on the `Run Simulation` then gets things going, and clicking on this button again pauses the simulation. If things are moving too slowly, or too quickly, pause the simulation and alter the speed through the speed slider as desired and then restart the simulation by clicking on the `Run Simulation` button again. The `view updates` setting (on `ticks` or `continuous`) can also be changed whenever desired (see Figure 3.1). Notice the difference in the speed of the turtle movements when this is done.

3.7.4 Exploratory exercises

For the first two sets of exercises, leave the sliders in their original settings (see Figure 3.25 and the discussion on sliders in the beginning of Section 3.7.2):

1. Perform several simulations for each question listed below—make a note of the placements of the predators and prey in the initial canvas *before each simulation*. For each simulation, once a pattern emerges that appears to provide an answer to the question posed, pause the simulation and make notes along the way.
   (a) Is it ever the case that either the predator or the prey (or both) die out? If yes, which of the two die out and under what circumstances (in terms of the initial placement of the predators and prey) does this appear to happen?
   (b) In general, under what circumstances do neither the predator nor the prey die out?

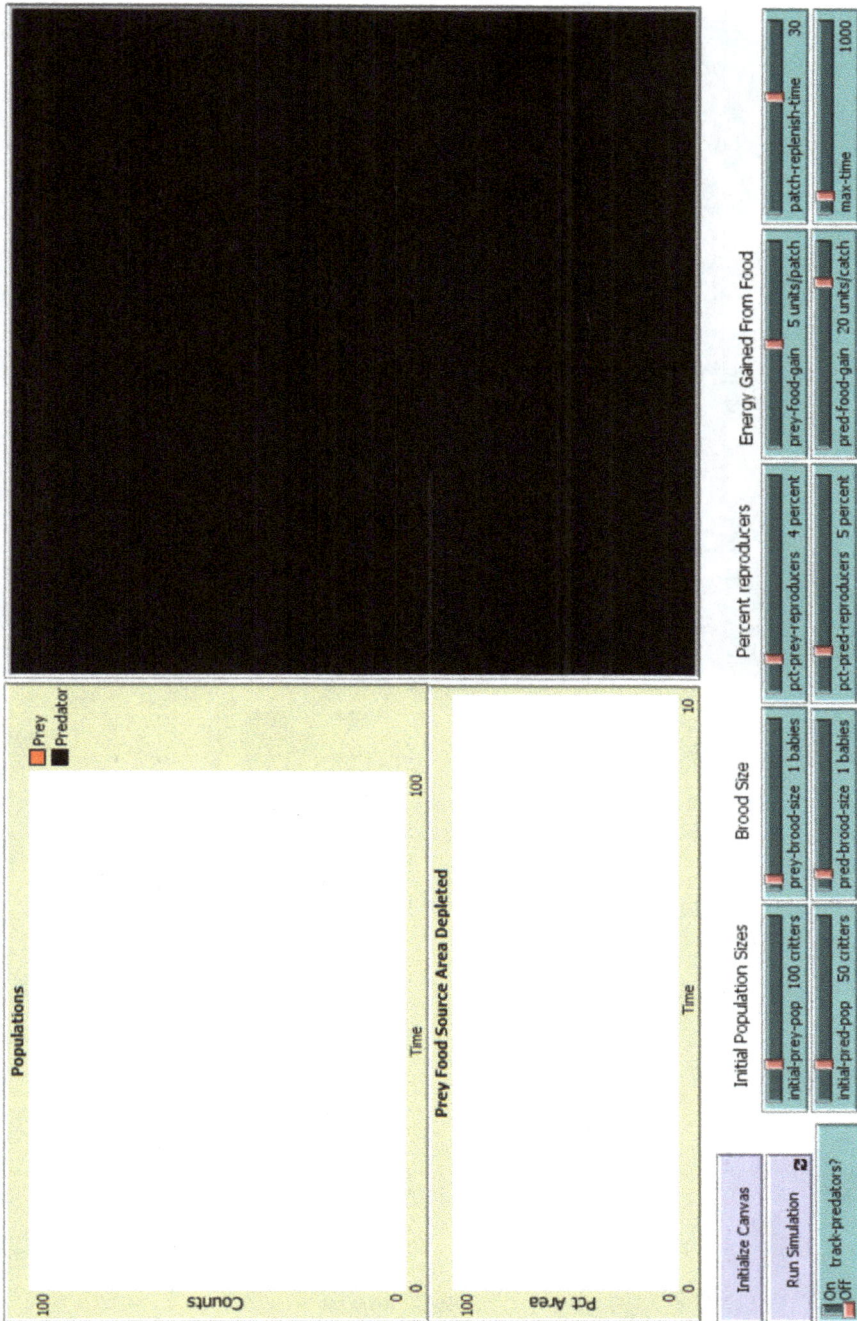

Figure 3.25: A possible starting interface window for the predator–prey model.

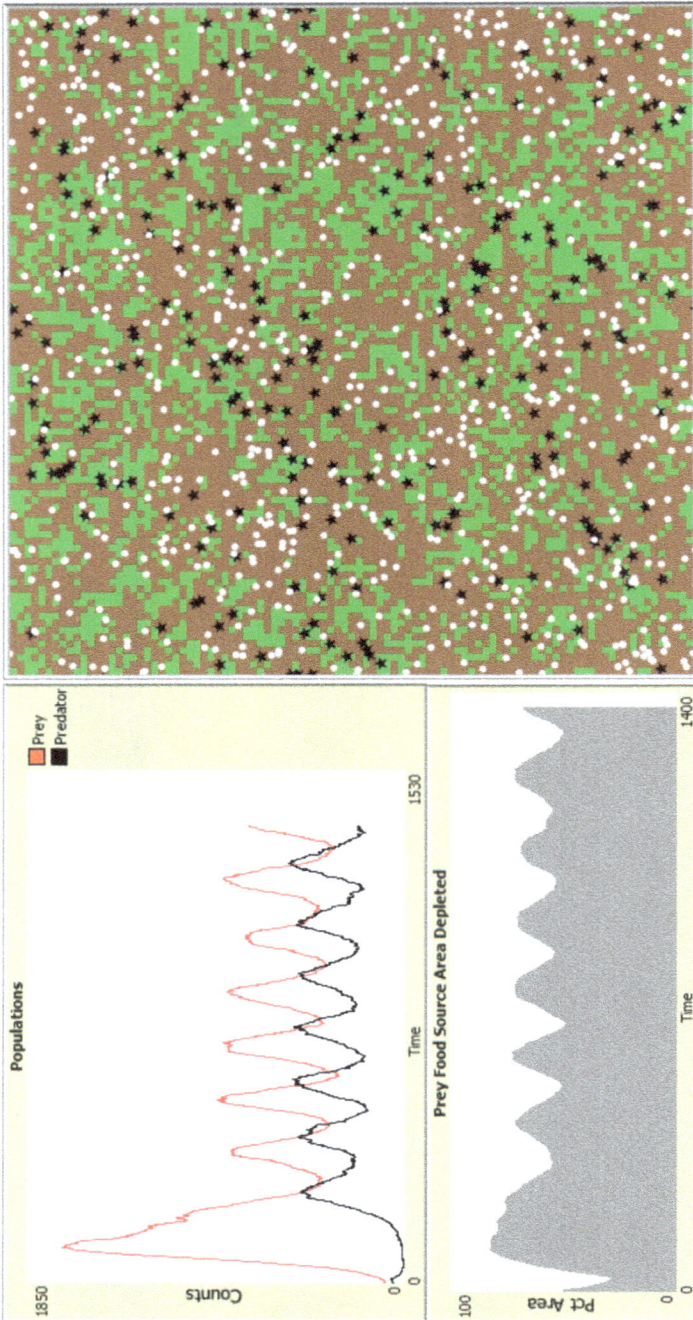

Figure 3.26: A sample simulation. First initialize the canvas, then run a simulation. The simulation can be stopped at any time by clicking on the Run Simulation button a second time.

(c) When neither the predator nor the prey die out, what generally appears to happen in the long run?

(d) When neither the predator nor the prey die out, what appears to be the pattern that the plots of predator and prey counts settle into? Again, make a note of the initial canvas appearance before each simulation.

(e) When neither the predator nor the prey die out, does there appear to be an approximate pattern that the predator and prey counts follow? If yes, what is this?

(f) Focusing on only the plot of prey counts and the plot of the percentage of area depleted of prey food sources (brown patches), does a connection appear to be present? If yes, what is this?

2. Create a duplicate copy of the original model, and call it PP2.nlog. Then right-click on the plot of the species population counts against time steps and select Edit to change this plot to one of predator counts against prey counts as shown in Figure 3.27. Keeping the original slider settings, try several simulations, focusing on only those in which neither the predator nor the prey die out. The simulations may be sped up or slowed down as desired.

(a) What appears to happen in the long run? What might the real-world explanation for this be?

(b) Does the plot ever settle on an equilibrium point? Explain, thinking in the real-world sense.

3. Reopen the original PP.nlogo model for the following. Play around with the prey brood-size slider (with all others in their original settings). Use the continuous view updates for some fairly spectacular simulation displays.

(a) Does anything interesting appear to happen as the prey brood-size is increased? If yes, explain.

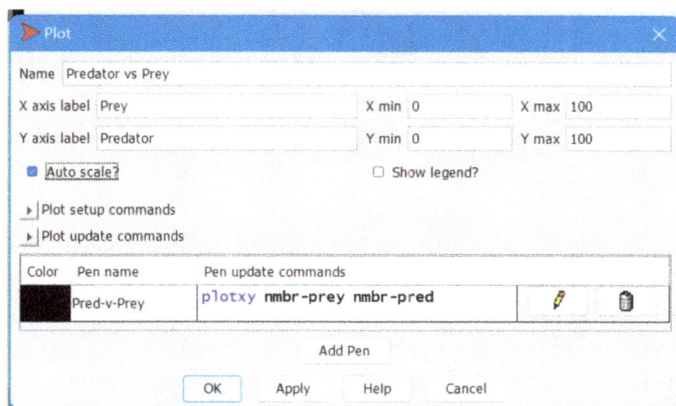

Figure 3.27: Settings for the predator against prey plot for Exercise 2.

(b) Is there a brood size that seems to consistently result in the prey eventually dying out? If yes, what seems to be the lowest such brood-size?
4. With the prey brood-size set at 4, play around with the predator brood-size slider (with all others in their original settings). Again, use the continuous view updates.
   (a) Does anything interesting happen as the predator brood-size is increased? If yes, explain.
   (b) Which combination(s) of brood-sizes seem to consistently result in neither of the species dying out in the long term?
5. Pick a pair of brood-sizes (from the previous simulations) that consistently results in a prey die-out and set the brood-size slides on these. Next, play around with the predator food-gain slider.
   (a) Does anything interesting come up? If yes, explain.
   (b) If the answer to the previous question is yes, what might a real-world explanation of this be?
6. Play around with more sliders in the PP.nlog model. Which settings (pairs, triples, or more) ensure in general that a species die-out *does not* occur? What are these?
7. Try playing around with the sliders in the PP2.nlogo model. Do the stabilizing settings (pairs, triples, or more) found in the previous exercise produce interesting patterns in the plots here?
8. Recall the "world wrap" features; see Figure 3.17. Uncheck the world wrap two boxes, then repeat the explorations conducted thus far. How does this tweak change things? Would this result in a more realistic view of things?
9. Here is an interesting coding exercise. In the procedures reproduce-prey and reproduce-pred (see Figure 3.21), the energy is shared equally between the parent and its offspring through the code

```
set energy-left (energy-left / 2)
```

This seems reasonable for brood-sizes of one. However, is it reasonable for larger brood-sizes? If no,
   (a) How might the code be tweaked to allow for such a "what-if" scenario?
   (b) What do simulations with the tweaked code reveal?

3.7.5 Extensions and other ideas

As mentioned previously in Section 2.5.5, there is much more to a true predator–prey system and actually following the two species in the wild to identify and record factors that influence their movements and/or behaviors and/or other characteristics of interest is at best difficult. However, one can hypothesize various "what-if" scenarios, and agent-based modeling (with the help of NetLogo) can be used to explore the possibilities.

The predator–prey model presented here may be modified in certain ways to account for additional characteristics of the system being modeled, or may be adapted to a different scenario. As previously mentioned, it is always helpful to pose questions before making changes. For example:

- Would it make sense to bring species gender into the picture? If no, why? If yes, why and how might this be accomplished?
- Would it make sense to bring age of individuals within one or both of the species into the picture? If no, why? If yes, why and how might this be accomplished?
- Would it make sense to make individuals' movements within the prey species dependent on what lies ahead? For example, if the new patch contains food or not. Or, if the new patch contains a predator. If no, why? If yes, why and how might this be accomplished?
- Would it make sense to make individuals' movements within the predator species dependent on what lies ahead? For example, if the new patch contains prey or not. Or, if the new patch contains a another member of its own species. If no, why? If yes, why and how might this be accomplished?
- If any one or more of the previous four scenarios is followed, what might an appropriate flowchart look like?
- If any one or more of the four scenarios is followed, what additional attributes will be needed? How might these be tracked?
- Are there further attributes that can be assigned to patches that will improve the realism of simulations? What are these, and how might they be incorporated into the model?
- What might be useful/informative simulation plots?

Again, as before, once a model is constructed there are other questions that can be asked. For example:

- For which parameters should sliders be included? Why?
- What information does varying each parameter provide?
- What is a meaningful goal of performing an analysis of the model through parameter variations?

Ignoring the free movement of species in a system and as hinted at in the questions just posed, here are some ways in which the previous predator–prey model might be extended or adapted to a different scenario.

- **Density-dependence**: Density-dependence on the rate at which the prey population size grows (or decreases) has been incorporated in the current model, after a fashion. How is this done? Can this be refined? If yes, how?
- **Dependence on age**: As for the previously seen compartmental model, it could be asked whether the age (of the prey and/or the predator) might play a role in the effectiveness of the predator pursuing and catching a prey. How might this idea be incorporated into the agent-based predator–prey model just constructed? How might agent attributes through turtles-own help facilitate this?

- **A two competing-species model**: How might the predator–prey example model just constructed be adapted to simulate a predator-free system in which two species compete with each other for a common food-source?
- **A competitor for the prey food-source**: How might the predator–prey model just constructed be adapted to include a second species that competes with the prey for the prey's food source? Suppose this (additional) species is not preyed upon by the predator, how might this scenario be modeled?
- **A competing predator species**: How might the predator–prey example model be adapted to so that it includes a second predator species that competes for the single prey?

There are other combinations that could be considered, and other questions that can be asked. For example:
- How can the loss of habitat be incorporated into wildlife population modeling?
- How can human encroachment be incorporated?
- How can invasive species be brought into the picture?

A useful outcome of asking all of these questions is that the modeling process can be initiated by first visualizing what goes on in a piece-by-piece manner before moving on to the big picture. Also, there are plenty of resources to fall back on in the event of encountering a brick wall.

3.8 Selected resources

Agent-based modeling is not subjected to many of the restrictions imposed on deterministic models such as the previously encountered compartmental models. For this reason, agent-based models can be applied to a much wider range of scenarios and systems. Here, a selection of resources on NetLogo and ideas for applications of agent-based models associated with mathematical biology or ecology are listed.

3.8.1 NetLogo

Here are some useful links specifically geared toward the use of NetLogo for agent-based modeling.

NetLogo User Community Models:
This page is contained on the NetLogo website at

> https://ccl.northwestern.edu/netlogo/models/community/index.cgi

This page contains models submitted in descending chronological order.

NetLogo Modeling Commons:
The link to this useful resource is

https://ccl.northwestern.edu/netlogo/docs/modelingcommons.html

NetLogo Resources and Links Page:
Located on the NetLogo website at

https://ccl.northwestern.edu/netlogo/resources.shtml

this page contains links to a wide range of NetLogo-specific resources beyond those already mentioned.

3.8.2 Background and theory

The epidemiology and population ecology resources listed in Section 2.6.2 remain relevant as underlying theory resources. The focus here is primarily on resources that directly relate to agent-based modeling, or that may contribute to developing such models.

An introduction to the practice of ecological modeling (2000) [65]
Brief description: Ecological modeling is one area for which agent-based modeling is ideally suited. The content of this article is presented with the new modeler in mind and, for this reason, does not adhere to a fixed sequence of processes. In the introductory comments to the article, the authors state

> We map one possible route through the sorts of decisions that will most likely need to be considered; this course is derived from our individual experiences plus the collective knowledge of our reviewers. We begin with conceptual models because many people, even self-labeled non-modelers, formulate conceptual models.

This article approaches ecological modeling more or less from first principles. It provides illustrations of lines of thought via flowcharts and outlines best practices in developing quantitative models. It also identifies possible pitfalls that may be encountered along the way. For these points alone, this article can serve as a useful (and brief) introduction for newcomers to the field, even though it does not specifically address agent-based models and/or Netlogo.

Individual-based modeling and ecology (2005) [52]
Brief description: This book is among the first published works that provides an in-depth treatment of individual-based modeling and its use to develop a theoretical understanding of how ecological systems work.

The approach introduced in this text, which the authors refer to as "individual-based ecology," lends itself directly to agent-based modeling. Even though NetLogo does not feature in the content, it is an excellent resource for establishing a solid foundation in the application of agent-based models to ecology.

Introduction to modeling in wildlife and resource conservation (2009) [96]
Brief description: The description for this book states

> This book provides students with the skills to develop their own models for application in conservation biology and wildlife management. Assuming no special mathematical expertise, the computational models used are kept simple and show how to develop models in both spreadsheet and programming language format.

The key point in this description is that no special mathematical expertise is assumed.

While the presentation in this book is not geared specifically toward agent-based modeling, the underlying theory provided through first principles will be of value to those entering into wildlife and resource conservation modeling.

The ODD protocol for describing agent-based and other simulation models:
A second update to improve clarity, replication, and structural realism (2020) [54]
Abstract: The Overview, Design concepts and Details (ODD) protocol for describing Individual- and Agent-Based Models (ABMs) is now widely accepted and used to document such models in journal articles. As a standardized document for providing a consistent, logical and readable account of the structure and dynamics of ABMs, some research groups also find it useful as a workflow for model design. Even so, there are still limitations to ODD that obstruct its more widespread adoption. Such limitations are discussed and addressed in this paper: the limited availability of guidance on how to use ODD; the length of ODD documents; limitations of ODD for highly complex models; lack of sufficient details of many ODDs to enable reimplementation without access to the model code; and the lack of provision for sections in the document structure covering model design rationale, the model's underlying narrative, and the means by which the model's fitness for purpose is evaluated. We document the steps we have taken to provide better guidance on: structuring complex ODDs and an ODD summary for inclusion in a journal article (with full details in Supplementary Material; Table 1); using ODD to point readers to relevant sections of the model code; update the document structure to include sections on model rationale and evaluation. We also further advocate the need for standard descriptions of simulation experiments and argue that ODD can in principle be used for any type of simulation model. Thereby ODD would provide a lingua franca for simulation modeling.

Notes: Along with the first 2006 publication of this protocol (see [50]) and the first update published in 2010 (see [51]), this article along with its included supplements should be considered a must read for agent-based modelers since it provides very useful information on strategies on how to design, construct, and use agent-based models effectively. Also of value is the reasoning behind the recommended changes to the originally proposed ODD protocol along with further suggestions provided in [51]. The listed references in all three publications may also provide ideas for projects under consideration. Another related article that will be of interest is [53], which provides a demonstration of the value of using the protocol to replicate agent-based models and produce new results with a replicated model. Needless to say, it would be a good idea to keep an eye out for further updates to this protocol.

Ecological modeling: A common-sense approach to theory and practice (2011) [47]

Brief description: This book examines four phases of the modeling process: conceptual model formulation, quantitative model specification, model evaluation, and model use. It provides useful building blocks for constructing systems simulation models and includes a format for reporting the development and use of simulation models.

By offering an integrated systems perspective to ecological modeling, this book can serve as a practical guide for students, teachers, and professional ecologists.

Models for planning wildlife conservation in large landscapes (2011) [90]

Brief description: This book introduces the field through a blend of conceptual, methodological, and application chapters with discussions on research, applications and concepts of modeling, and ideas and strategies for wildlife habitat models used in conservation planning.

This book may serve as a single-resource volume of information on the most current and effective techniques of modeling for wildlife conservation in large landscapes (at least up through 2011). It is described as being appropriate for students and researchers alike.

An introduction to agent-based modeling: Modeling natural, social, and engineered complex systems with NetLogo (2015) [122]

Brief description: Written by the developer of NetLogo, this book is a comprehensive and hands-on introduction to the core concepts, methods, and applications of agent-based modeling, including NetLogo examples.

Agent-based modeling in population studies (2017) [55]

Brief description: This book features the work of leading experts who share their unique insights into their experience with agent-based modeling. Focusing on the use of agent-based modeling in population studies from concepts to applications, it also discusses best practices to future developments and provides a solid point of reference for those wanting to use agent-based models in population research.

How new concepts become universal scientific approaches: Insights from citation network analysis of agent-based complex systems science (2018) [117]

Abstract: Progress in understanding and managing complex systems comprised of decision-making agents, such as cells, organisms, ecosystems, or societies, is like many scientific endeavours limited by disciplinary boundaries. These boundaries, however, are moving and can actively be made porous or even disappear. To study this process, I advanced an original bibliometric approach based on network analysis to track and understand the development of the model-based science of agent-based complex systems (ACS). I analysed research citations between the two communities devoted to ACS research, namely agent-based (ABM) and individual-based modeling (IBM). Both terms refer to the same approach, yet the former is preferred in engineering and social sciences, while the latter prevails in natural sciences. This situation provided a unique case study for grasping how a new concept evolves distinctly across scientific domains and how to foster convergence into a universal scientific approach. The present analysis based on novel hetero-citation metrics revealed the historical development of ABM and IBM, confirmed their past disjointedness, and detected their progressive merger. The separation between these synonymous disciplines had silently opposed the free flow of knowledge among ACS practitioners and thereby hindered the transfer of methodological advances and the emergence of general systems theories. A surprisingly small number of key publications sparked the ongoing fusion between ABM and IBM research. Beside reviews raising awareness of broad-spectrum issues, generic protocols for model formulation and boundary-transcending inference strategies were critical means of science integration. Accessible broad-spectrum software similarly contributed to this change. From the modeling viewpoint, the discovery of the unification of ABM and IBM demonstrates that a wide variety of systems substantiate the premise of ACS research that microscale behaviours of agents and system-level dynamics are inseparably bound.

Notes: There are two useful features of this article. First, the analysis of the literature conducted by the author provides support to the coalescing of the terms individual-based and agent-based modeling. The discussion leading to this conclusion is quite interesting. Second, the references listed for this article contain quite a few articles that might find use in a search for ideas.

Agent-based models (2019) [44]

Brief description: This book is intended for readers who are new to agent-based modeling. In it, the author considers a range of methodological and theoretical issues; shows how to design an agent-based model with a simple example; offers some practical advice about developing, verifying and validating agent-based models; and finally discusses how to plan an agent-based modeling project, publish the results and apply agent-based modeling to formulate and evaluate social and economic policies. The author also offers a brief, but thorough, treatment of a cutting-edge technique; practical advice about how to design and create agent-based models; and includes carefully chosen examples from different disciplines.

Agent-based and individual-based modeling: A practical introduction (2019) [101]

Brief description Described by the authors as a continuation of their previous book, titled *Individual-based modeling and ecology*, [52], this book is an excellent resource for gaining a strong foundation in agent-based modeling *and* the use of NetLogo.

Agent-based model history and development (2019) [46]

Abstract: Agent-based modeling has a deep rich history. When it began in physics in the 1930s, it immediately resulted in key scientific breakthroughs. Through time, many disciplines both in and outside academia have adopted agent-based modelling for scientific investigation, especially where systems made up of people were concerned. All this makes it an ideal tool with which to investigate the economy.

Notes: Some may find this article to be a useful historical resource with ideas on how to extend applications of agent-based models outside of traditional areas of use.

The history of agent-based modeling in the social sciences (2021) [103]

Abstract: Agent-based modeling is a powerful technique that allows modeling social phenomena ab initio or from first principles. In this paper, we review the history of agent-based models and their role in the social sciences. We review 62 papers and create a timeline of developments starting from 1759 and Adam Smith into the recent past of 2020 and efforts to model the Covid-19 pandemic. We reflect on model validation, different levels of model complexity, multi-scale models, and cognitive architectures. We identify key trends for the future use of agent-based modeling in the socials sciences.

Notes: This article provides a window into the social scientists point of view. It is very likely that ideas explored may extend to other areas of interest.

Agent-based modeling in the philosophy of science (2023) [38]

Abstract: Agent-based models (ABMs) are computational models that simulate behavior of individual agents in order to study emergent phenomena at the level of the community. Depending on the application, agents may represent humans, institutions, microorganisms, and so forth. The agents' actions are based on autonomous decision-making and other behavioral traits, implemented through formal rules. By simulating decentralized local interactions among agents, as well as interactions between agents and their environment, ABMs enable us to observe complex population-level phenomena in a controlled and gradual manner.

This entry focuses on the applications of agent-based modeling in the philosophy of science, specifically within the realm of formal social epistemology of science. The questions examined through these models are typically of direct relevance to philosophical discussions concerning social aspects of scientific inquiry.

Notes: This article also provides a window into the social scientists point of view of applications of agent-based models. Again, ideas explored may motivate extensions to areas of interest outside of the social sciences.

Using the ODD protocol and NetLogo to replicate agent-based models (2025) [53]

Abstract: Replicating existing models and their key results not only adds credibility to the original work, it also allows modelers to start model development from an existing approach rather than from scratch. New theory can then be developed by changing the assumptions or scenarios tested, or by carrying out more in-depth analysis of the model. However, model replication can be challenging if the original model description is incomplete or ambiguous. Here, we show that the use of standards can facilitate and speed up replication: the ODD protocol for describing models, and NetLogo, an easy-to-learn but powerful software platform and language for implementing agent-based models. To demonstrate the benefits of this approach, we conducted a replication experiment on 18 agent-based models from different disciplines. The researchers doing the replications had no or little previous experience using ODD and NetLogo. Their task was to rewrite the original model description using ODD, implement the model in NetLogo, and try to replicate at least one exemplary main result. They were also asked to produce, if time allowed, some initial new results with the replicated model, and to record the total time spent on the replication exercise. Replication was successful for 15 out of 18 models. The time taken varied between 2 and 12 days, with an average of 5 days. ODD helped to systematically scan the original model description, while NetLogo proved easy and quick to learn, but difficult to debug when implementation problems arose. Although most of the models replicated were relatively simple, we conclude that even for more complex models it can be useful to use ODD and NetLogo for replication, at least for developing a prototype to help decide how to proceed with the replicated model. Overall, the use of both, standard approaches such as ODD and easy to learn but powerful software such as NetLogo, can promote coherence and efficiency within and between different models and modeling communities. Imagine if all modelers spoke ODD and NetLogo as a common language or *lingua franca*.

Notes: This article provides a further discussion on and explorations of the value of using NetLogo and the previously mentioned ODD protocol for agent-based modeling.

Agent-based modeling resource website [3]
Brief description: This website of the Columbia University Mailman School of Public Health contains lists of books, articles, links to other websites, and courses addressing background information for modeling in medicine in general, and applications of agent-based modeling to medicine.

3.8.3 Research and ideas

The articles listed here are in chronological order and, in addition to abstracts for each article listed, some brief notes are provided. These notes include comments on why the article in question was chosen as well as its possible use as a resource. This may involve possible direct applications of the method(s) discussed as well as possible extensions to equivalent or analogous projects.

Dynamic models of segregation (1971) [106]

Abstract: Some segregation results from the practices of organizations, some from specialized communication systems, some from correlation with a variable that is non-random, and some results from the interplay of individual choices. This is an abstract study of the interactive dynamics of discriminatory individual choices. One model is a simulation in which individual members of two recognizable groups distribute themselves in neighborhoods defined by reference to their own locations. A second model is analytic and deals with compartmented space. A final section applies the analytics to neighborhood tipping. The systemic effects are found to be overwhelming: there is no simple correspondence of individual incentive to collective results. Exaggerated separation and patterning result from the dynamics of movement. Inferences about individual motives can usually not be drawn from aggregate patterns. Some unexpected phenomena, like density and vacancy, are generated. A general theory of 'tipping' begins to emerge.

Notes: This is considered to be among the earliest (formal) applications of the agent-based approach to modeling; see also [48] and [120].

Ten years of individual-based modeling in ecology: What have we learned and what could we learn in the future? (1999) [49]

Abstract: Each modeler who builds and analyzes an individual-based model learns of course a great deal, but what has ecology as a whole learned from the individual-based models published during the last decade? Answering this question proves extremely difficult as there is no common motivation behind individual-based models. The distinction is introduced between 'pragmatic' motivation, which uses the individual-based approach as a tool without any reference to the theoretical issues which have emerged from the classical state variable approach and 'paradigmatic' motivation, which explicitly refers to theoretical ecology. A mini-review of 50 individual-based animal population models shows that the majority are driven by pragmatic motivation. Most models are very complex and special techniques to cope with this complexity during their analysis are only occasionally applied. It is suggested that in order to orient individual-based modeling more toward general theoretical issues, we need increased explicit reference to theoretical ecology and an advanced strategy for building and analyzing individual-based models. To this end, a heuristic list of rules is presented, which may help us to advance the practice of individual-based modeling and to learn more general lessons from individual-based modeling in the future than we have during the last decade. The main ideas behind these rules are as follows: (1) Individual-based models usually make more realistic assumptions than state variable models, but it should not be forgotten that the aim of individual-based modeling is not 'realism' but modeling. (2) The individual-based approach is a bottom-up approach which starts with the 'parts' (i. e., individuals) of a system (i. e., population) and then tries to understand how the system's properties emerge from the interaction among these parts. However, bottom-up approaches alone will never lead to theories at the systems level. State variable or top-down approaches are needed to provide an appropriate integrated view, i. e., the relevant questions at the population level.

Notes: While this mini-review of the literature through 1999 emphasizes individual-based models in the title and the exposition, it is worth mentioning that the word "individual" could just as easily be replaced by the word "agent" (see [117]). It should also be noted that the articles listed in the bibliography and references for this review most likely did not involve the use of NetLogo since all were published before 1999 after which NetLogo became readily available; see the FAQs in [121].

The author classifies the motivations behind the publications reviewed as being "pragmatic" (those that emphasize the attitude that individual-based modeling is a new tool in the ecological modelers' toolbox), or "paradigmatic" (those that emphasize the attitude that something might be wrong with the classical theory). The examples illustrating these two motivations along with a collection of heuristic rules provides a look into some useful ways of thinking. Finally, the bibliography and references listed may serve as a useful resource for ideas.

Empirically based, agent-based models (2006) [67]

Abstract: There is an increasing drive to combine agent-based models with empirical methods. An overview is provided of the various empirical methods that are used for different kinds of questions. Four categories of empirical approaches are identified in which agent-based models have been empirically tested: case studies, stylized facts, role-playing games, and laboratory experiments. We discuss how these different types of empirical studies can be combined. The various ways empirical techniques are used illustrate the main challenges of contemporary social sciences: (1) how to develop models that are generalizable and still applicable in specific cases, and (2) how to scale up the processes of interactions of a few agents to interactions among many agents.

Notes: This article provides useful information for those who might be interested in combining empirical methods with agent-based modeling. While this article is intended for social scientists, the ideas and methods discussed quite naturally extend beyond social sciences.

Agent-based land-use models: A review of applications (2007) [86]

Abstract: Agent-based modeling is an approach that has been receiving attention by the land use modeling community in recent years, mainly because it offers a way of incorporating the influence of human decision-making on land use in a mechanistic, formal, and spatially explicit way, taking into account social interaction, adaptation, and decision-making at different levels. Specific advantages of agent-based models include their ability to model individual decision-making entities and their interactions, to incorporate social processes and non-monetary influences on decision-making, and to dynamically link social and environmental processes. A number of such models are now beginning to appear it is timely, therefore, to review the uses to which agent-based land use models have been put so far, and to discuss some of the relevant lessons learned, also drawing on those from other areas of simulation modeling, in relation to future applications. In this paper, we review applications of agent-based land use models under the headings of (a) policy analysis and planning, (b) participatory modeling, (c) explaining spatial patterns of land use or settlement, (d) testing social science concepts and (e) explaining land use functions. The greatest use of such models so far has been by the research community as tools for organizing knowledge from empirical studies, and for exploring theoretical aspects of particular systems. However, there is a need to demonstrate that such models are able to solve problems in the real world better than traditional modeling approaches. It is concluded that in terms of decision support, agent-based land-use models are probably more useful as research tools to develop an underlying knowledge base which can then be developed together with end-users into simple rules-of-thumb, rather than as operational decision support tools.

Notes: For those intending on pursuing studies that focus on assessing the impacts of land use (by humans) on natural systems, this article provides background information and a further example of combining empirical methods with agent-based models.

Agent-based modeling of animal movement (2010) [113]

Abstract: Animal movement is a complex spatiotemporal phenomenon that has intrigued researchers from many disciplines. Interactions among animals, and between animals and the environments that they traverse play an important role in the development of the complex ecological and social systems in which they are embedded. Agent-based models have been increasingly applied as a computational approach to the study of animal movement across landscapes. In this article, we present a review of agent-based models in which the simulation of animal movement processes and patterns is the central theme. Our discussion of these processes is focused on four key components: internal states, external factors, motion capacities, and navigation capacities. These four components have been identified in the emerging movement ecology research paradigm and are important for modeling animal movement behavior. Because agent-based models allow for an individual-based approach that encapsulates these four components, the underlying processes that drive animal behavior can be deeply explored using this technique. A set of challenges and issues remain, however, for agent-based models of animal movement. In this article, we review the existing literature and identify potential research directions that could help address these challenges.

Notes: The model presented in this article might provide ideas for a different migration model. Or, it might provide ideas for modeling an unrelated phenomenon that closely mimics migration behavior.

An introduction to agent-based models as an accessible surrogate to field-based research and teaching (2010) [93]

Abstract: There are many barriers to fieldwork including cost, time, and physical ability. Unfortunately, these barriers disproportionately affect minority communities and create a disparity in access to fieldwork in the natural sciences. Travel restrictions, concerns about our carbon footprint, and the global lockdown have extended this barrier to fieldwork across the community and led to increased anxiety about gaps in productivity, especially among graduate students and early-career researchers. In this paper, we discuss agent-based modeling as an open-source, accessible, and inclusive resource to substitute for lost fieldwork during COVID-19 and for future scenarios of travel restrictions such as climate change and economic downturn. We describe the benefits of Agent-Based models as a teaching and training resource for students across education levels. We discuss how and why educators and research scientists can implement them with examples from the literature on how agent-based models can be applied broadly across life science research. We aim to amplify awareness and adoption of this technique to broaden the diversity and size of the agent-based modeling community in ecology and evolutionary research. Finally, we discuss the challenges facing agent-based modeling and discuss how quantitative ecology can work in tandem with traditional field ecology to improve both methods.

Notes: Instructors at institutions with limited resources who wish to simulate fieldwork in their course work will find this article helpful.

Agent-based inference for animal movement and selection (2010) [59]

Abstract: Contemporary ecologists often find themselves with an overwhelming amount of data to analyze. For example, it is now possible to collect nearly continuous spatiotemporal data on animal locations via global positioning systems and other satellite telemetry technology. In addition, there is a wealth of readily available environmental data via geographic information systems and remote sensing. We present a modeling framework that utilizes these forms of data and builds on previous research pertaining to the quantitative analysis of animal movement. This approach provides additional insight into the environmental drivers of residence and movement as well as resource selection while accommodating path uncertainty. The methods are demonstrated in an application involving mule deer movement in the La Sal Range, Utah, USA. Supplemental materials for this article are available online.

Notes: This article may provide ideas for similar studies involving the movement of different species.

The role of agent-based models in wildlife ecology and management (2011) [88]

Abstract: Conservation planning of critical habitats for wildlife species at risk is a priority topic that requires the knowledge of how animals select and use their habitat, and how they respond to future developmental changes in their environment. This paper explores the role of a habitat-modeling methodological approach, agent-based modeling, which we advocate as a promising approach for ecological research. Agent-based models (ABMs) are capable of simultaneously distinguishing animal densities from habitat quality, can explicitly represent the environment and its dynamism, can accommodate spatial patterns of inter- and intra-species mechanisms, and can explore feedbacks and adaptations inherent in these systems. ABMs comprise autonomous, individual entities; each with dynamic, adaptive behaviors and heterogeneous characteristics that interact with each other and with their environment. These interactions result in emergent outcomes that can be used to quantitatively examine critical habitats from the individual- to population-level. ABMs can also explore how wildlife will respond to potential changes in environmental conditions, since they can readily incorporate adaptive animal-movement ecology in a changing landscape. This paper describes the necessary elements of an ABM developed specifically for understanding wildlife habitat selection, reviews the current empirical literature on ABMs in wildlife ecology and management, and evaluates the current and future roles these ABMs can play, specifically with regards to scenario planning of designated critical habitats.

Notes: This article may be a source of ideas for creating and applying models to any number of wildlife ecology and management problems.

Incorporating behavioral ecological strategies in pattern-oriented modeling of caribou habitat use in a highly industrialized landscape (2012) [107]

Abstract: Woodland caribou (*Rangifer tarandus*) are classified as threatened in Canada, and the Little Smoky herd in west-central Alberta is at immediate risk of extirpation due in part, to anthropogenic activities such as oil, gas, and forestry that have altered the ecosystem dynamics. Winter season represents an especially challenging time of year for this Holarctic species as it is characterized by a shortage of basic resources and is when most industrial development occurs, to which caribou can perceive as increased predation risk. To investigate the impact of industrial features on caribou, we developed a spatially explicit, agent-based model (ABM) to simulate the underlying behavioral mechanisms caribou are most likely to employ when navigating their landscape in winter. The ABM model is composed of cognitive caribou agents possessing memory and decision-making heuristics that act to optimize tradeoffs between energy acquisition and predator/disturbance avoidance. A set of environmental data layers was used to develop a virtual grid representing the landscape in terms of forage availability, energy content, and predation-risk. The model was calibrated with caribou bio-energetic values from literature sources, and validated using GPS data from thirteen caribou radio-collars deployed over 6 months from 2004 to 2005. Simulations were conducted on alternative caribou habitat-selection strategies by assigning different fitness-maximizing goals to agents. The model outcomes were evaluated using a pattern-oriented modeling approach with actual caribou data. The scenario in which the caribou agent must trade off the mutually competing goals of obtaining its daily energy requirement, conserving reproductive energy, and minimizing predation risk, was found to be the best-fit scenario. Not recognizing industrial features as risk causes simulated caribou to unrealistically reduce their daily and landscape movements; equally, having risk take precedence results in unrealistic energetic deficits and large-scale movement patterns, unlike those observed in actual caribou. These results elucidate the most likely behavioral strategies caribou use to select their winter habitat, the relative extent to which they perceive industry features as potential predation, and the differential energetic costs associated with each strategy. They can assist future studies of how caribou may respond to continued industrial development and/or mitigation measures.

Notes: This article may provide ideas on creating models that address the impact of land use on any number of migratory (or nonmigratory) species.

Individual-based models in ecology after four decades (2014) [36]

Abstract: Individual-based models simulate populations and communities by following individuals and their properties. They have been used in ecology for more than four decades, with their use and ubiquity in ecology growing rapidly in the last two decades. Individual-based models have been used for many applied or pragmatic issues, such as informing the protection and management of particular populations in specific locations, but their use in addressing theoretical questions has also grown rapidly, recently helping us to understand how the sets of traits of individual organisms influence the assembly of communities and food webs. Individual-based models will play an increasingly important role in questions posed by complex ecological systems.

Notes: The authors revisit the ideas of "pragmatic" and "paradigmatic" motivated models, introduced in the previously mentioned article [49]. This article presents a review and discussions of articles covering a wide range of applications in ecology, so it contains a list of interesting applications of agent-based models to ecology—in the beginning of the article the authors mention that the terms individual-based and agent-based are interchangeable.

Modeling tiger population and territory dynamics using an agent-based approach (2015) [28]

Abstract: Effective conservation planning of globally endangered tigers (*Panthera tigris*) requires a good understanding of their population dynamics. Territoriality, an essential characteristic of many wildlife species, plays a crucial role in the population dynamics of tigers. However, previous models of tiger population dynamics have not adequately incorporated territoriality. We therefore developed and implemented a spatially explicit agent-based model of tiger population dynamics shaped by different territorial behaviors of males and females. To allow for predictions to new conditions, for which no data exist, territories are not imposed but emerge from the tigers' perception of habitat quality and from their interactions with each other. Tiger population dynamics is deduced from merging territory dynamics with observed demographic rates. We apply the model to Nepal's Chitwan National Park, part of a global biodiversity hotspot and home to a large (~ 125) population of tigers. Our model matched closely with observed patterns of the real tiger population in the park, including reproduction, mortality, dispersal, resource selection, male and female land tenure, territory size and spatial distribution, and tiger population size and age structure. The ultimate purpose of the model, which will be presented in follow-up work, is to explore human-tiger interactions and assess threats to tiger populations across contexts and scales. The model can thus be used to better inform decision makers on how to conserve tigers under uncertain and changing future conditions.

Notes: This article may be useful to someone who might be interested in developing and implementing a spatially explicit agent-based model for some other cat (or other predator) species.

Making predictions in a changing world: The benefits of individual-based ecology (2015) [111]

Abstract: Ecologists urgently need a better ability to predict how environmental change affects biodiversity. We examine individual-based ecology (IBE), a research paradigm that promises better a predictive ability by using individual-based models (IBMs) to represent ecological dynamics as arising from how individuals interact with their environment and with each other. A key advantage of IBMs is that the basis for prediction fitness maximization by individual organisms is more general and reliable than the empirical relationships that other models depend on. Case studies illustrate the usefulness and predictive success of long-term IBE programs. The pioneering programs had three phases: conceptualization, implementation, and diversification Continued validation of models runs throughout these phases. The breakthroughs that make IBE more productive include standards for describing and validating IBMs, improved and standardized theory for individual traits and behavior, software tools, and generalized instead of system-specific IBMs. We provide guidelines for pursuing IBE and a vision for future IBE research.

Notes: This article provides useful insights, and guidelines, for pursuing individual-based ecology through agent-based modeling.

An agent-based modeling approach to determine winter survival rates of American robins and eastern bluebirds (2016) [64]

Abstract: American robins (*Turdus migratorius*) and Eastern bluebirds (*Sialia sialis*) are two species of migratory thrushes that breed in Northwest Indiana, but historically are uncommon during the winter season. These trends have changed recently, and both species are seen more abundantly during the winter. Recently invaded non-native fruiting plants continue to provide nutrients for the birds throughout the winter and may contribute to the increased avian populations during that time. To measure the effect these food sources contribute to thrush wintering habits, we created an agent-based computer model to simulate the birds' movement in Northwest Indiana, along with their food consumption over the course of the winter season. The model incorporates availability of food sources, foraging and roosting behavior, bio-energetics, and starvation, with parameter values informed by the literature. We obtained simulated winter survival rates of the birds that could begin to explain the changes in the birds' migratory patterns.

Notes: This article provides a clear illustration of using the ODD protocol for constructing an agent-based model. It also includes a flowchart that illustrates how periods of sleep may be incorporated into a model.

Modeling the spread of the Zika virus at the 2016 Olympics (2017) [85]

Abstract: The Zika virus is an arbovirus that is spread by mosquitoes of the Aedes genus and causes mild fever-like symptoms. It is strongly associated with microcephaly, a condition that affects development of fetal brains. With the recent emergence of Zika in Brazil, we develop an agent-based model to track mosquitoes and humans throughout the 2016 Olympics in Rio de Janeiro to investigate how the Olympics might affect the spread of the virus. There are many unknowns regarding the spread and prevalence of Zika, with approximately 80 % of infected individuals unaware of their infectious status. We therefore discuss results of experiments where several unknown parameters were varied, including the rate at which mosquitoes successfully bite humans, the percentage of initially infected mosquitoes, and the sizes of the human and mosquito populations. From these experiments, we make initial predictions regarding effective control measures for the spread of Zika.

Notes: Just as in the previous article, this article also provides a clear illustration of using the ODD protocol in constructing an agent-based model. It also provides an example of constructing flowcharts for complex systems.

Agent-based models of the Green Dot Bystander Violence Prevention Program on college campuses (2017) [70]

Abstract: Despite the hard work of a diverse collection of organizations committed to violence prevention, the prevalence of rape, abuse, and other forms of interpersonal violence remains startling, especially on college campuses. Here, we present an agent-based model (ABM) of interpersonal violence rooted in the philosophy of the Green Dot Bystander Training Program, in the hopes of providing insight into ways in which training of students can be improved so that intervention attempts are more effective. Two models, with and without adaptive behaviors, are studied under two population sizes. Through sensitivity testing, various outcomes are analyzed to measure the effectiveness of each intervention strategy. The scenarios that result in the smallest relative number of violent acts are those with a denser population, while the adaptive models produce unexpected results that prompt questions about human behavior and our tendency toward bystander intervention.

Notes: This article addresses a social science question. However, the approach used to explore the issue might provide ideas for developing analogous models applicable to biological or ecological scenarios.

Agent-based model of salmon migration in the Snake River (2017) [87]

Abstract: Agent-based modeling in biological systems is a relatively new approach that has been used increasingly in examining environments that are difficult to describe through traditional statistical models. Studies in ecology have been particularly eager to employ this technique as the emphasis on the agents allow for studying conservation efforts with measurable effects on the population as a whole. In this paper, we showcase an agent based model of the process of salmon returning to their freshwater spawning ground as they pass through the lower Snake River. Based on a model originally developed by Amy Steimke, we have updated the model to encompass more recent fish run data and provide a statistical analysis of the model that was not in Steimke's original project. In the process of working with the model, we produced a distribution that suggested that the dam objects were causing the salmon population to spread out, a dimension of difficulties to salmon reproduction that is not captured purely in the threat to life that dams pose. After quantifying this observed spread, we developed an addition to the model for opportunity-based reproduction to capture the effects of such a spread on the reproductive success of the salmon population. We find that the submodel results suggest a clear impact of dams upon the reproductive success. This significant finding leads us to conclude with potential implications and expansions to the work presented here.

Notes: This article may be helpful for conservation biologists engaged in inland fisheries studies.

Decision-making in agent-based modeling: A current review and future prospectus (2019) [35]

Abstract: All basic processes of ecological populations involve decisions; when and where to move, when and what to eat, and whether to fight or flee. Yet decisions and the underlying principles of decision-making have been difficult to integrate into the classical population-level models of ecology. Certainly, there is a long history of modeling individuals' searching behavior, diet selection, or conflict dynamics within social interactions. When all the individuals are given certain simple rules to govern their decision-making processes, the resultant population level models have yielded important generalizations and theory. But it is also recognized that such models do not represent the way real individuals decide on actions. Factors that influence a decision include the organism's environment with its dynamic rewards and risks, the complex internal state of the organism, and its imperfect knowledge of the environment. In the case of animals, it may also involve complex social factors, and experience and learning, which vary among individuals. The way that all factors are weighed and processed to lead to decisions is a major area of behavioral theory.

While classic population-level modeling is limited in its ability to integrate decision-making in its actual complexity, the development of individual- or agent-based models (IBM/ABMs) (we use ABM throughout to designate both "agent-based modeling" and an "agent-based model") has opened the possibility of describing the way that decisions are made, and their effects, in minute detail. Over the years, these models have increased in size and complexity. Current ABMs can simulate thousands of individuals in realistic environments, and with highly detailed internal physiology, perception, and ability to process the perceptions and make decisions based on those and their internal states. The implementation of decision-making in ABMs ranges from fairly simple to highly complex; the process of an individual deciding on an action can occur through the use of logical and simple (if-then) rules to more sophisticated neural networks and genetic algorithms. The purpose of this paper is to give an overview of the ways in which decisions are integrated into a variety of ABMs and to give a prospectus on the future of modeling of decisions in ABMs.

Notes: This broad review of agent-based models and their applications provides useful reading for those interested in taking advantage of the contributions that agent-based modeling can make to decision-making.

Modeling neural behavior and pain during bladder distention using an agent-based model of the central nucleus of the amygdala (2019) [12]

Abstract: Chronic bladder pain evokes asymmetric behavior in neurons across the left and right hemispheres of the amygdala. An agent-based computational model was created to simulate the firing of neurons over time and in response to painful bladder stimulation. Each agent represents one neuron and is characterized by its location in the amygdala and response type (excited or inhibited). At each time step, the firing rates (Hz) of all neurons are stochastically updated from probability distributions estimated from data collected in laboratory experiments. A damage accumulation model tracks the damage accrued by neurons during long-term, painful bladder stimulation. Emergent model output uses neural activity to measure temporal changes in pain attributed to bladder stimulation. Simulations demonstrate the model's ability to capture acute and chronic pain and its potential to predict changes in pain similar to those observed in the lab. Asymmetric neural activity during the progression of chronic pain is examined using model output and a sensitivity analysis.

Notes: This article may provide ideas on applying agent-based modeling to medicine.

An overview of agent-based models in plant biology and ecology (2020) [127]

Abstract: Agent-based modeling (ABM) has become an established methodology in many areas of biology, ranging from the cellular to the ecological population and community levels. In plant science, two different scales have predominated in their use of ABM.

One is the scale of populations and communities, through the modeling of collections of agents representing individual plants, interacting with each other and with the environment. The other is the scale of the individual plant, through the modeling, by functional structural plant models (FSPMs), of agents representing plant building blocks, or metamers, to describe the development of plant architecture and functions within individual plants. The purpose of this review is to show key results and parallels in ABM for growth, mortality, carbon allocation, competition, and reproduction across the scales from the plant organ to populations and communities on a range of spatial scales to the whole landscape. Several areas of application of ABMs are reviewed, showing that some issues are addressed by both population-level ABMs and FSPMs. Continued increase in the relevance of ABM to environmental science and management will be helped by greater integration of ABMs across these two scales.

Notes: Plant biology and ecology can also be studied using agent-based models. This overview provides a useful review of background information that will help in this area.

An investigation of mitigation measures on the spread of COVID-19 in a college classroom using agent-based modeling (2023) [112]

Abstract: In this manuscript, we describe the process of using agent-based modeling in NetLogo to create a simulation of COVID-19 spread in a traditional college classroom. The model allows for an evaluation of different preventative measures implemented by the University of Pittsburgh, including the cohort classroom attendance model, mask and vaccine mandates, contact tracing, and classroom sanitation. Through the use of the model's interactive interface, the impact of adjusting specific measures by the institution could be visualized, providing a valuable tool for combating diseases that spread through droplet transmission.

Notes: The approach used in this article can be adapted to analogous scenarios involving different infectious diseases.

4 Self-organizing maps

4.1 Introduction

This chapter expands on the presentation in [8], which introduces an open-source app prepared specifically to perform data clustering, a method used in the analysis of high-dimensional data. The app implements a clustering algorithm using *Shiny* [29], an R package with which interactive web apps can be built using R, [100].

The chapter begins with an overview of clustering and the clustering algorithm used by the app as described in [8]. This is followed by directions on getting access to, and setting up the app. Then the key steps in using the app are outlined. The remainder of this chapter follows the structure of the previous chapters—two examples of using the app are given with sample exploratory exercises and ideas for extensions. Finally, a selection of further resources on self-organizing maps and cluster analysis is provided.

4.2 Self-organizing maps and clustering

Informally, a *cluster* refers to a group of observations within a larger data set that exhibit similar attributes, and *clustering* refers to the actual process of partitioning the whole data set into reasonably homogeneous groups. The clustering process itself uses an algorithm that takes advantage of a similarity measure of some form to classify observations in a data set in a manner that permits the grouping.

4.2.1 The general idea

Since clustering algorithms classify data observations into classes based on attributes present in the data, and without direction from the analyst, these algorithms are referred to as *unsupervised classification algorithms* (as opposed to *supervised classification algorithms*). While there are several such unsupervised classification algorithms (see, e. g., [6]), the app under discussion uses what is called a *self-organizing map* (commonly abbreviated as SOM).

As mentioned in [6], a self-organizing map, also called a *Kohonen map*, makes use of *artificial neural networks* to produce a low-dimensional graphical representation of high-dimensional data. This is done in a way that not only preserves the structure in the original data, but also brings to light any similarities present in the data by providing information that can be used to construct visual aids to identify clusters within the data. The process by which a self-organizing map accomplishes a clustering of data in a high-dimensional data set can be summarized as follows.

In the simplest of terms, a self-organizing map takes *input nodes*, observations from the data, and maps these to a collection of *output nodes*, which provide the information

https://doi.org/10.1515/9783111609560-004

needed to produce a low-dimensional graphical representation of clusters present in the data. The process by which this is accomplished involves the use of what is referred to as a *competitive learning algorithm*.

4.2.2 The competitive learning algorithm

Suppose the data comprise n rows of observations with columns associated with a collection of $p \geq 3$ variables. Corresponding to each row, $i = 1, 2, \ldots, n$ of the data, let the vector $\mathbf{x}_i = (x_{i1}, x_{i2}, \ldots, x_{ip})$ contain the preferably scaled non-dimensional entries of the ith row in the data. The index i identifies an input node.

Stage 1—Initialization
A set of units representing the output nodes are typically arranged in a square lattice of dimensions at most $\lfloor \sqrt{n} \rfloor \times \lfloor \sqrt{n} \rfloor$. Then the connection between each input node i and every output node j is represented by a *weight vector* $\mathbf{w}_j = (w_{j1}, w_{j2}, \ldots, w_{jp})$ having entries from the interval $[0, 1]$ assigned either randomly or using prior knowledge of the data set.

Note: From a biological point of view, the output nodes in a self-organizing map can be associated with *neurons* and the components of the vectors $\mathbf{w}_j$ can be associated with *synapses*. These form the artificial neural network mentioned earlier. However, these terms will not always be used in what follows. An input node will sometimes be referred to as such, or as a data observation, and an output node will sometimes be referred to as an output node, or simply a node when this is obvious.

After the initialization of the weight vectors, the algorithm begins a sequence of iterative processes.

Stage 2—Training the weight vectors
To start this stage, a data observation (an input node), say $\mathbf{x}_i$, is selected at random.

Finding the best matching unit: The similarity of $\mathbf{x}_i$ to each of the weight vectors $\mathbf{w}_j$ is measured using a distance function, commonly the Euclidean distance. That is, for the chosen input node i and for each output node j, the distance

$$d(\mathbf{x}_i, \mathbf{w}_j) = \sqrt{\sum_{k=1}^{p} (x_{ik} - w_{jk})^2}$$

is calculated. The weight vector that minimizes this distance is called the *winning weight vector*, or *best matching unit* for $\mathbf{x}_i$. Let J denote the best matching unit for the input node i, that is,

$$J = \arg \min_j \{d(\mathbf{x}_i, \mathbf{w}_j)\}.$$

Then, once the best matching unit J for the input node i is found, a process referred to as *activation* takes place for all weight vectors within some predetermined neighborhood of the best matching unit. This leads to the *competitive learning* part of the algorithm.

Training the weight vectors: Denote the predetermined neighborhood of the best matching unit by $N(J)$ and define

$$h_{Jj} = \begin{cases} \eta(t), & j \in N(J), \\ 0, & j \notin N(J), \end{cases}$$

where the monotonically decreasing function $\eta(t)$ defines the learning rate. See, for example, [6] for further details on defining neighborhoods and on options available for learning rate functions.

For the Kohonen som function, which the app uses, the learning rate decreases linearly from 0.05 to 0.01 and the neighborhood decreases linearly from 2/3 of the nearest nodes on the map to zero. When the neighborhood is 1 or less, only the winning weight vector learns. Hence, for this app, the self-organizing map starts with a cooperative and competitive learning process, then switches to a strictly competitive learning process.

Now, let $t = 1, 2, \ldots, T$ denote time steps, with $t = 1$ being the starting time-step. Then the current weight vectors, denoted $\mathbf{w}_j(1)$, are updated (they are trained) using the predetermined number, T, of iterations of the activation function using

$$\mathbf{w}_j(t+1) = \mathbf{w}_j(t) + h_{Jj}[\mathbf{x}_i - \mathbf{w}_j(t)].$$

Once all of the weight vectors in the neighborhood of the node J have been updated, or trained, another different input node is selected at random and Stage 2 is repeated. The process ends once all input nodes have been selected, and all of the weight vectors have been trained for every input node.

Stage 3—Visualization
At the termination of the iterations in Stage 2, there are available the input nodes, the output nodes arranged in a square lattice, and the trained weight vectors connecting each input node to output nodes.

It is worth noting that the trained weight vectors themselves will be partitioned into $K \leq n$ clusters. It is this information that is used to project graphical representations of similarities as clusters in the data onto the two-dimensional lattice of output nodes.

4.3 RStudio

These days it is not uncommon to have high-dimensional massive data sets in research projects, and finding similarities among attributes for such data sets without the use of advanced software is at best very difficult. In fact, the use of advanced software in multivariate data analysis is a necessity and reducing the computational cost to analyze data has become vitally important. However, software is typically expensive and the coding involved can be time consuming. This is where R, and the previously mentioned Shiny app can prove useful, particularly in an educational setting.

R, a free software environment for statistical computing and graphics, will need to be downloaded from the *R Project for Statistical computing* website at

https://www.r-project.org/

The downloading and installation process is surprisingly simple and quick. For general purposes, many prefer using *RStudio* [104], a popular coding environment that runs R and that has several helpful user-interface features. This is also the preferred environment in which to run the app in question. RStudio is also free and should be downloaded from

https://posit.co/downloads/

The downloading and installation instructions for this are also quite straightforward. It will be noticed in the RStudio installation instructions that R (commonly referred to as *base R*) has to be loaded and installed prior to downloading and installing RStudio.

The Shiny self-organizing map app, which is best run in RStudio, provides a user interface for the som function contained in package kohonen; see [27] and [78]. Since the app provides the user interface to perform the needed clustering process, no prior knowledge of coding in R is needed. For this reason, this chapter (and this book) does not dwell on any instructions on coding in R. The emphasis is on only acquiring and setting up the app and then using it. For those interested, a selection of resources on R, RStudio, and Shiny are provided in Section 4.8.1.

4.4 The self-organizing map app

Download this app from the *Intercollegiate Biomathematics Alliance* GitHub platform, located at

https://github.com/iba-community/R-Shiny-for-Clustering

To download the needed files, click on the green Code button and then select Download ZIP.

Next, create a folder named, for example, ShinySomApp and extract the files from the downloaded zipped folder. Then move the contents out of the resulting R-Shiny-for-Clustering-main folder into the ShinySomApp folder and delete the empty R-Shiny-for-Clustering-main folder. There are some preliminary steps before easy access to the app can be achieved.

4.4.1 Running the app

When RStudio is started up three windows show up (see Figure 4.1), the *Console* being on the left, one showing the *Global Environment* in the upper right, and the third showing the *Home* folder in the lower right. The first step is to find the Shiny app folder just created. From the Files shown in the lower-right window, locate and click on the ShinySomApp folder. The next step is to install certain packages needed by the app—*this has to be done only once for each computer.*

To do this, open the requirements.R file shown in the lower right-hand window by clicking on it. This file will then appear in a fourth window just above the console on the left (see Figure 4.2). Run the code by clicking on the Source button that appears in the top-right side of the requirements.R window. After some busywork, the R Console will show the ">" symbol. This means the needed R packages have been successfully downloaded. To get rid of console clutter, select Clear Console from the Edit drop-down menu and then close the requirements.R file.

The app can now be started up by choosing the file Main.R in the lower right-hand window, and then clicking on the Run App button that appears in the top-right side of the Main.R window (see Figure 4.3). When this is done, the app opens in its own window. A slightly cropped part of the opening window for the app has the appearance of Figure 4.4.

4.4.2 App preliminaries

Once started up, as shown in Figure 4.4, there are three tabs in the app that navigate the user through the three main windows.

Introduction:
This tab opens the Introduction window, which is what the app starts up in. Here, the user is introduced to the purpose of this application, given brief directions on how to use it, and some sample plots that it produces.

Import Data:
This tab opens the Import Data window; see Figure 4.5. This is where users are given the option to upload their own data by selecting the option Upload Data, or explore the app with data provided in the app under Import a Sample Dataset (Section 4.5 provides an example of this).

Note:
Should the user wish to upload their own data, then the data must be placed in, and imported from a *.csv file in which rows represent observations and columns represent

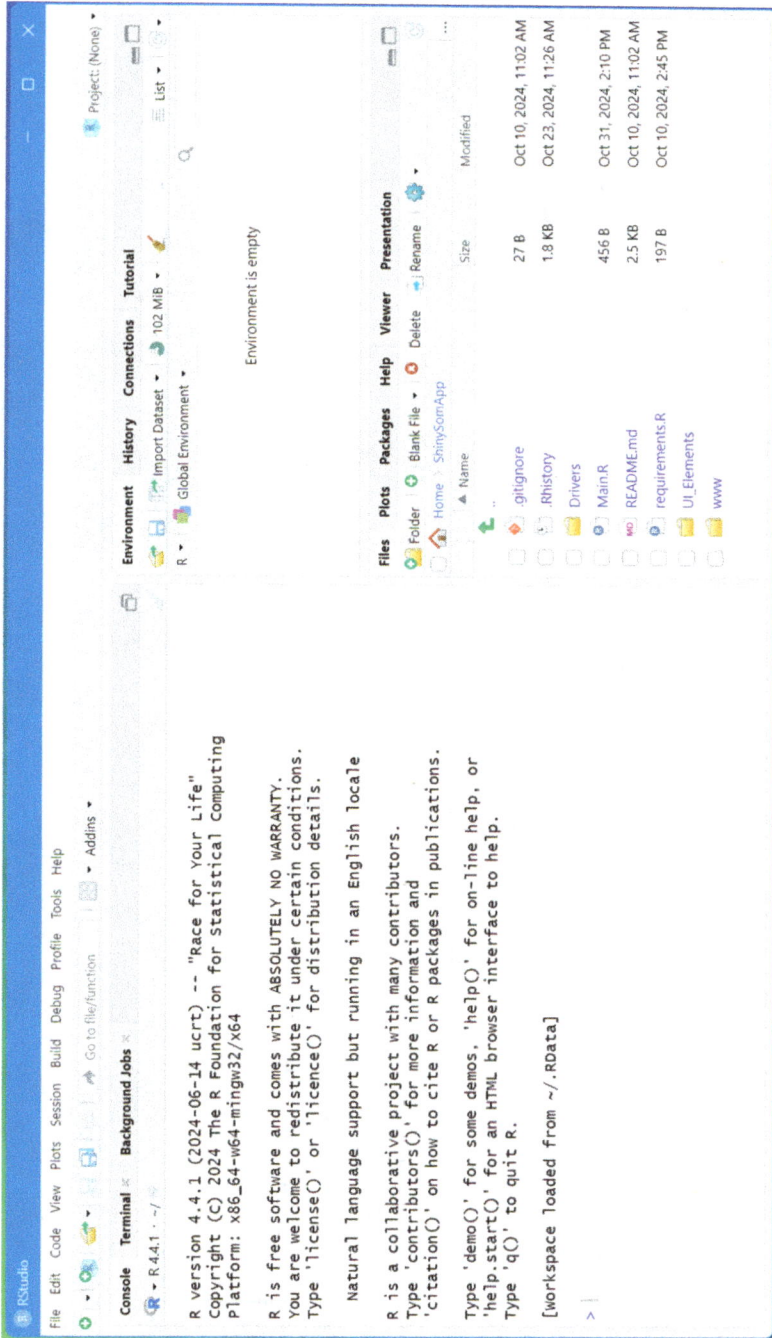

Figure 4.1: The opening RStudio window with the ShinySomApp folder opened.

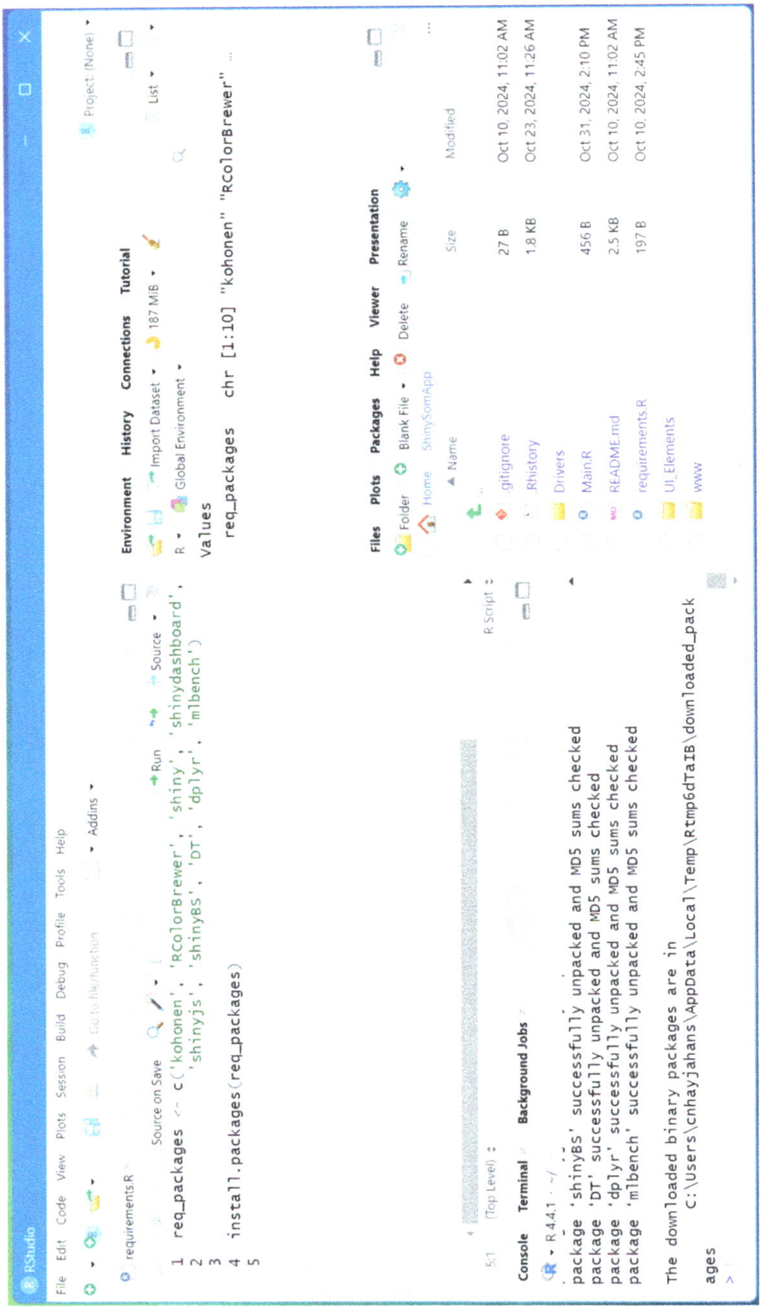

Figure 4.2: The `requirements.R` script file contains code to install needed packages. To run the code, click on the `Source` button. Once the packages have been loaded, they will be listed in the upper right-hand `Global Environment` window and will also be listed under `Packages` in the lower right-hand window.

RStudio

File Edit Code View Plots Session Build Debug Profile Tools Help

Addins ▾

Go to file/function

Project: (None) ▾

Main.R ×

Run App ▾

```
1   # Import Libraries
2   library(shiny)
3   library(shinydashboard)
4   library(shinydashboardPlus)
5   library(shinyjs)
6   library(shinyBS)
7   library(DT)
8   library(dplyr)
9
10  # Import R Scripts
11  source("UI_Elements/UI_Sidebar.R")
12  source("UI_Elements/UI_Body.R")
13  source("Drivers/Server.R")
14
15  # Create Header
16  header <- dashboardHeader(title = "Self Organizing Map")
```

6:17   (Top Level) ‡

R Script ‡

Environment   History   Connections   Tutorial

Import Dataset ▾   190 MiB ▾   List ▾

R ▾   Global Environment ▾

Values

req_packages   chr [1:10] "kohonen" "RColorBrewer" ...

Files   Plots   Packages   Help   Viewer   Presentation

Folder   Blank File ▾   Delete   Rename

Home ▸ ShinySomApp

▲ Name	Size	Modified
.gitignore	27 B	Oct 10, 2024, 11:02 AM
.Rhistory	1.8 KB	Oct 23, 2024, 11:26 AM
Drivers		
Main.R	456 B	Oct 31, 2024, 2:10 PM
README.md	2.5 KB	Oct 10, 2024, 11:02 AM
requirements.R	197 B	Oct 10, 2024, 2:45 PM
UI_Elements		
www		

Console   Terminal ×   Background Jobs ×

R ▾ R 4.4.1 · ~/

>

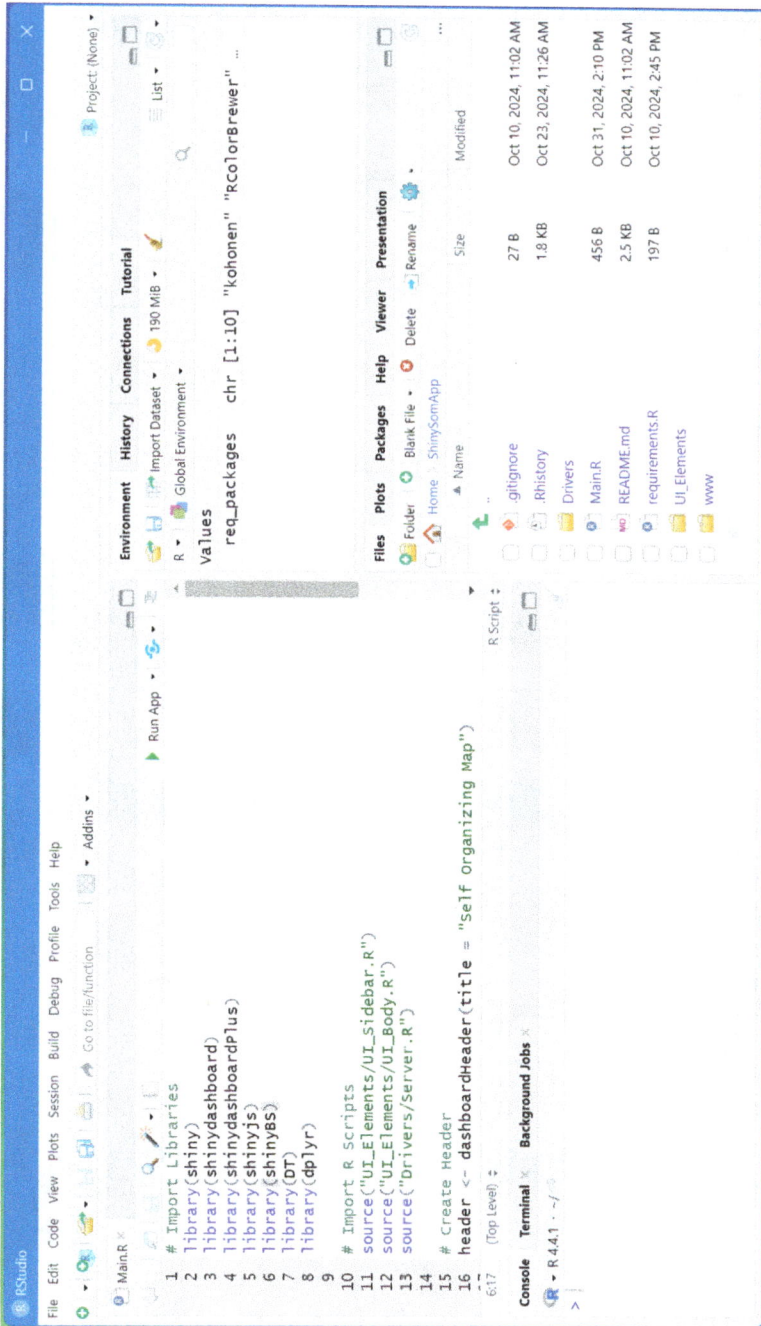

Figure 4.3: The `Main.R` script file contains code to start up and run the app. To run the code, click on the `Run App` button.

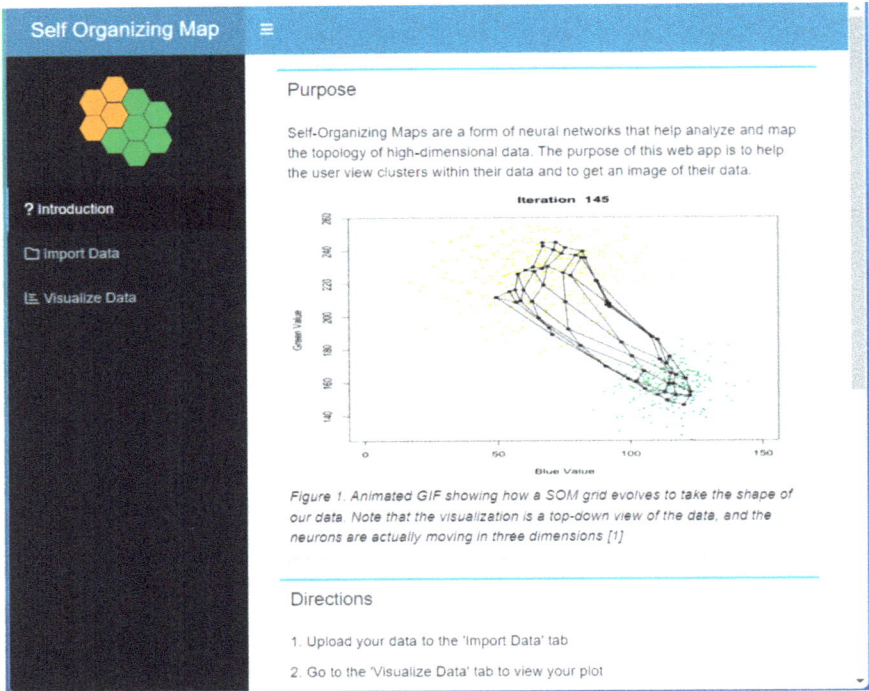

Figure 4.4: The Introduction window. Scroll down to see more.

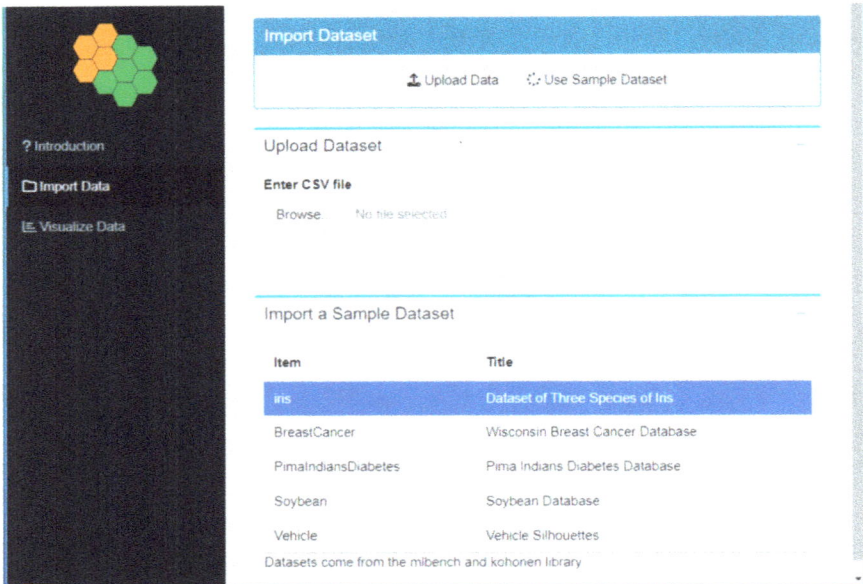

Figure 4.5: The Import Data window.

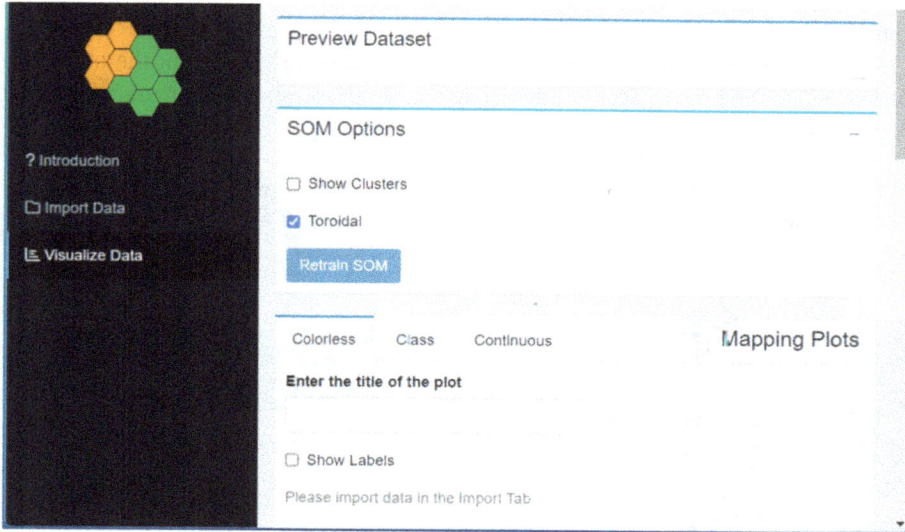

Figure 4.6: The Visualize Data window. Here, the user selects the variables of interest and chooses from the various self-organizing map parameter values and other options. Notice that the Toroidal box is checked by default.

variables for which the observations are made. The data should be checked for missing entries, and rows containing missing entries should be removed. Also, if a labels column is included then its entries should be character strings that are not too long (Section 4.7 provides an example of this).

Visualize Data:

This tab opens the Visualize Data window; see Figure 4.6. The graphics produced by the self-organizing map app are automatically displayed in this window. Illustrations of choices in this window appear in the example that follows.

4.5 Fisher's iris data

Fisher's iris data, [40], is one of the most common benchmark sample data sets that is used in the validation of clustering algorithms. This data set is used here; see, for example, [40] and [99] for more.

Very briefly, observations on 50 each of three species of the iris flower family, *iris setosa, iris virginica*, and *iris versicolor* are obtained ($n = 150$ observations total) under four variables. Data under these four variables are obtained through measurements of the *sepal length, sepal width, petal length*, and *petal width* for each flower. A fifth column in the data contains the labels variable, which identifies the species to the which each flower belongs.

Spoiler:

It turns out that the iris setosa species is clearly distinguishable from the other two species, while the iris versicolor and virginica are not as clearly distinguishable from each other.

To start things off, click on the Import Data tab, select iris from the Import a Sample Dataset option, and then click on the Use Sample Dataset button; see Figure 4.7.

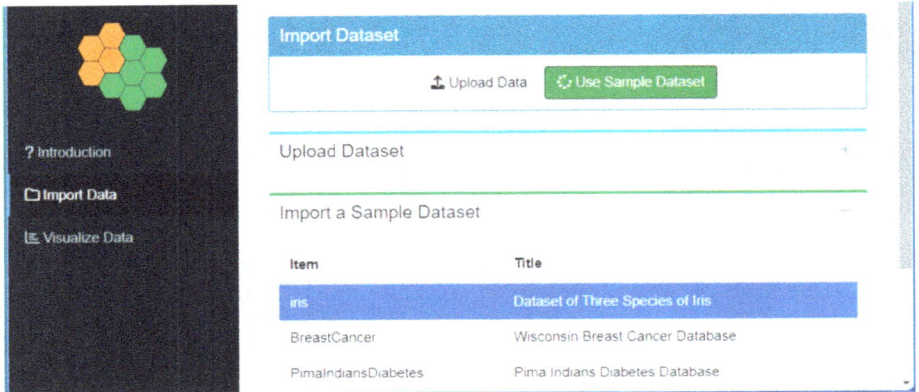

Figure 4.7: Notice that the Use Sample Dataset button is green because it has been clicked on, and the iris dataset is highlighted because it has been clicked on.

Next, click on the Visualize Data tab and select the variables of interest: sepal length, sepal width, petal length, and petal width in the SOM Options window by selecting the variables from the drop-down menu that opens when the Which columns do you want to use? box is clicked on (see Figure 4.8). Next, for this example, set the SOM Size to 5. This sets the dimensions of the square lattice for the nodes to 5×5. Different dimensions can be chosen, as long as they do not exceed $\lfloor \sqrt{n} \rfloor \times \lfloor \sqrt{n} \rfloor$.

Note that the variables can also be identified in the Preview Dataset window, see Figure 4.8, by clicking on an observation under each variable name (be aware that clicking on the variable name does NOT select the variable. Instead, clicking on the variable name reorganizes the data).

It is recommended that a self-organizing map should be *retrained* (see the button in Figure 4.6) several times before coming to any conclusions about the data. In [27], the randomness involved in the self-organizing map initialization process is provided as a rationale for doing this. To retrain a self-organizing map, click on the Retrain SOM button in the Visualize Data window. Plots are automatically updated every time a retraining is performed. Then plots of interest can be viewed, or downloaded with the Download button located at the bottom left of every graphic. Lastly, Toroidal is on by

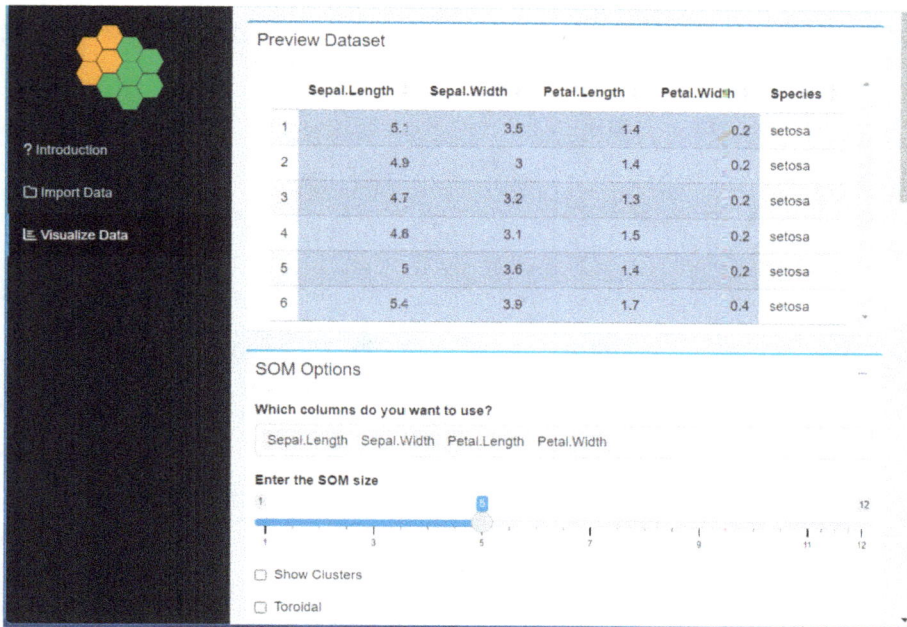

Figure 4.8: Here, the four variables are selected, the SOM size is set to 5, and the Show Clusters and Toroidal boxes are unchecked.

default. This turns maps into planar representations of a torus—this should be turned off for this example (uncheck the box, see Figure 4.8).

Scrolling down in the Visualize Data window (see Figure 4.8) reveals two types of plots with generic titles, *mapping plots* and *analysis plots*. Examples of versions of these plots with brief descriptions appear below.

4.5.1 Mapping plots

The focus of these plots is the separation of the data into clusters and one of three versions of mapping plots—*colorless*, *class*, or *continuous* plots—can be viewed.

Colorless mapping plot:
The *colorless mapping plot* shows observations from the data as dots contained in hexagons, the hexagons being the output nodes; see Figure 4.9. Observe that some of the nodes are associated with observations and some (the empty ones) are not. In general, observations associated with a node and with nearby nodes have similar weight vectors, indicating the corresponding data observations have similar attributes. If so desired, the generic plot title (Colorless Mapping Plot) can be replaced by entering an alternate title such as shown in Figure 4.9.

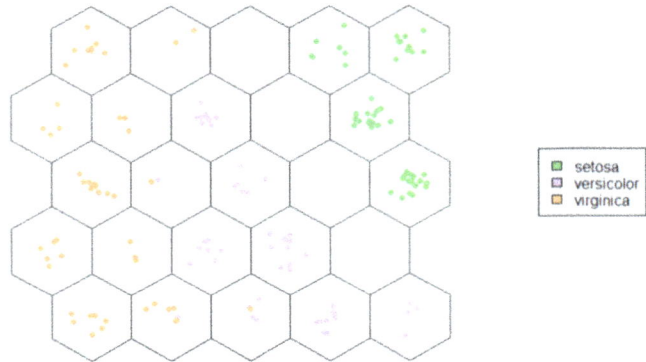

Colorless Class Continuous   ·

Enter the title of the plot

Colorless Mapping Plot of the Iris Data

☑ Show Labels

What column do you want to use to label the
dataset?

Species   ▾

Mapping Plots

Colorless Mapping Plot of the Iris Data

☐ setosa
☐ versicolor
☐ virginica

⬇ Download

Figure 4.9: The clearly isolated top-right cluster is of the species setosa. Virginica and versicolor, while
mostly separated, become partially mixed toward the center.

The self-organizing map algorithm is unsupervised, as are all clustering processes,
so the som function does not require labels for the colorless map plot. However, labels
are useful and can be added afterwards if a labels variable is available in the data. In
fact, the colorless mapping plot becomes more informative if a labels variable for the
data is available. For example, the iris data has *Species* as the labels variable in the last
column of the data set; see Figure 4.8. To add labels (and colors) to the colorless mapping
plot, as well as a legend, click on Show Labels and select the variable Species from the
drop-down list. Then the mapping plot in Figure 4.9 identifies each observation by a
color associated with its corresponding species as indicated in the legend.

Figure 4.9 was obtained after retraining the map several times until the cluster of
iris setosa appeared clearly separated from the other two clusters of iris virginica and
iris versicolor. Following the position of the setosa cluster on the map after every retraining, it will be noticed that the setosa cluster will always be quite clearly separated

from the other, sometimes mixed, clusters, and will appear in a different corner of the map after each retraining.

Notice that for the current retraining, Figure 4.9 shows the setosa cluster in the top-right corner, the virginica cluster on the left side, and the versicolor cluster in the middle. The current separations of virginica and versicolor are reasonably clear, but further retrainings may result in clearer (or less clearer) separations/clustering.

If so desired, this plot can be downloaded as a *.png file by clicking on the Download button located in the bottom-left corner; see Figure 4.9.

Class representation plot:

The *class plot* (see Figure 4.10) requires a labels column in the data, which must be selected from the drop-down list as for the previous map. This plot follows directly from the Figure 4.9 version of the colorless mapping plot, also with the labels variable Species. Here, each of the nodes (again represented by hexagons) is colored according to the color associated with the most frequently occurring species tied to the node. As the included legend indicates, gray hexagons represent nodes that do not contain any observations.

What column do you want to use to label the dataset?

Species	▼

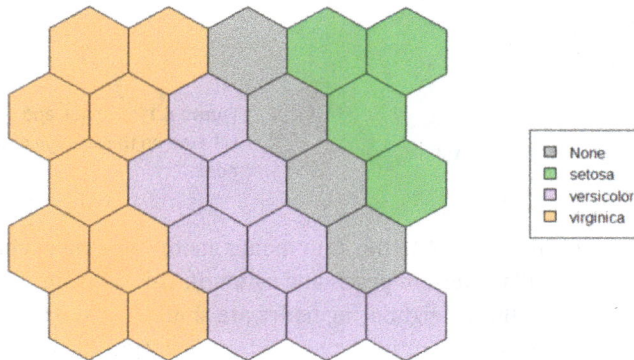

Figure 4.10: Unlike the colorless map in Figure 4.9, individual data cases are not shown and colors are assigned to whole nodes.

For comparison purposes, focus on the nodes containing both viginica and versicolor in Figure 4.9. The nodes containing a majority of one species in Figure 4.9 are given the color of the majority in Figure 4.10. For example, the second node from the left in the bottom row of Figure 4.9 contains more of the virginica species, so the corresponding node in Figure 4.10 has the color assigned to virginica. On the other hand, again in Figure 4.9, the node to the immediate right of this node has more of the versicola species, so the corresponding node in Figure 4.10 has the color assigned to versicolor. Finally, observe that the (empty) gray nodes form the clear separation of the setosa cluster from the virginica and versicolor mixed clusters.

Continuous response map:

The *continuous response map* (see Figure 4.11) uses *classical multidimensional scaling* to create the equivalent of a three-dimensional map to describe the clusters apparent in the class representation plot—the third dimension being represented by shades of similar or different hues. For this map, the average values of the weight vectors for the nodes are treated as RGB values and the node is colored according to its RGB value. Then the map displays clusters with similar hues, but varying shades. Different shades of a similar hue among a collection of nodes indicate nodes within the same cluster.

Continuous Response Map

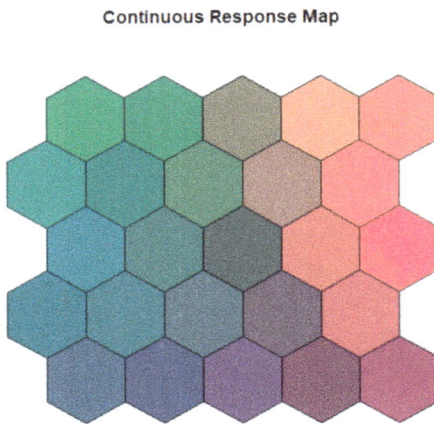

Figure 4.11: The hue and shade assigned to a node is based on the average of the weight vectors for the node.

For example, in Figure 4.11 the four nodes in the top right cluster (of setosa) are colored with similar hues (of peach and salmon). This indicates that the average of the weight vectors of these neighboring nodes are similar. On the other hand, the virginica and versicolor clusters mix hues from solid green (virginica, on the left) to bluish green and then to purple (versicolor, primarily in the middle to bottom right).

4.5.2 Analysis plots

These plots retain the clusters shown in Figures 4.9–4.11 with some additional information. As for mapping plots, one of three versions of analysis plots—*counts, neighborhood distance*, or *codes* plots—can be viewed.

Counts Plot:

In the *Counts Plot* nodes are represented by discs (see Figure 4.12) and each disc is colored according to the number of observations associated with the corresponding node. The legend for this plot displays a spectrum of colors with associated counts and gray discs represent nodes that are not associated with any data observations.

Counts   Neighbourhood Distance   Codes   Analysis Plots

Enter the title of the plot

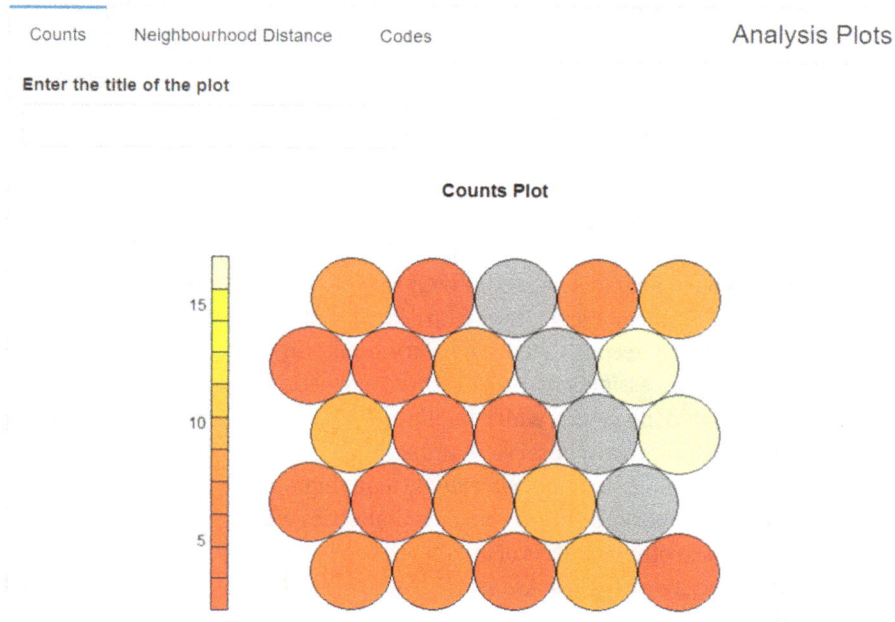

Counts Plot

Figure 4.12: The color assigned to each node represents the number of observations that lie within the node. Gray nodes are empty nodes.

Observe that in Figure 4.12 the gray discs clearly separate the setosa species from the other two. The setosa cluster is clumped together with nodes having moderate to high counts. The virginica cluster in the left of the map is more spread out with lower counts per node. Similarly, the versicolor cluster has about the same number of observations in each node as the virginica cluster.

Neighborhood distance plot:

The nodes in *Neighborhood Distance Plot* (see Figure 4.13) are also represented by discs. For this plot, the color assigned to a node provides a visual representation of the sum of the distances of the weights of the node in question to weights in its immediate neighboring nodes.

Neighborhood Distance Plot

Figure 4.13: Each node color provides a visual idea of how "far" (or different) its members are from its immediate neighbors through a distance measure.

Here, the legend displays a spectrum of colors with their corresponding sums of distances. This means that a node having a color that corresponds to a small sum of distances will tend to have neighboring nodes with similar weight vectors (similar neighbors), and a node having a color associated with a large sum of distances will have neighboring nodes with weight vectors that are not very similar (more dissimilar neighbors).

In Figure 4.13, colors assigned to the discs within clusters indicate that each of the clusters have low sums of distances, which implies that observations within neighboring nodes are similar. Alternatively, observe that the majority of the nodes *not* associated with *any* data observations have higher values, implying that neighboring nodes are (obviously) quite different. So, one may conclude that nodes within the setosa cluster are very likely close to each other in terms of attributes, but further away from observations belonging to the other two clusters. As for the other two, neighboring nodes associated with the virginica cluster are close together, but nodes associated with the versicolor cluster are moderately close to each other.

Codes plot:

The *Codes Plot* (see Figure 4.14) includes a feature that permits distinguishing characteristics within clusters apparent in the class representation plot. Here, nodes are represented by discs that contain different colored sectors of circles (wedge-shaped pieces) having different radii. Each wedge represents one of the variables of interest, identified by its color. The radius of the wedge is proportional to the average magnitude of weight vectors associated with the input vectors corresponding to the species represented by the node in question. Informally, the size of these wedges provide a measure of the contribution that each variable makes to the clustering process. The legend for this plot shows the contributing variables and their corresponding colors.

So, for example, the four top-right discs associated with the setosa cluster suggest that the variable that contributes almost entirely to the identification of this cluster is sepal width. Curiously, the five neighboring empty nodes also show wedges—because of the information provided by the counts plot, these can be ignored. Next, the codes for

Codes Plot

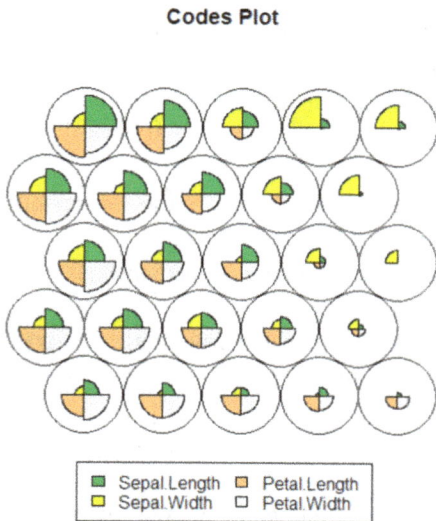

Figure 4.14: Each wedge provides a visual representation of a measure of the contribution that the corresponding variable makes to the node (and cluster) in question.

the virginica cluster (to the left) suggest that sepal length, petal length, and petal width all contribute approximately equally to the clustering process. Finally, the codes for the versicolor cluster suggest that petal length and petal width are the two that consistently contribute equally to the clustering.

4.5.3 Exploratory exercises

There is one caution that should be brought up before proceeding on to any exploratory exercises. This is that it is highly unlikely that plots arrived at by different individuals or different runs will be the same (or even similar) in appearance. This is because of the randomness built into the self-organizing mapping algorithm. However, with enough re-trainings all plots will eventually provide equivalent information about clusters present in the data.

Unlike the examples in the previous chapters, there are no parameters to play with here. But it might be interesting to determine whether the findings proposed in interpretations of Figures 4.9–4.14 can be generally replicated through a sequence of retraining, and hence, are reasonable. For example:
1. Can a clearly separated cluster for setosa always be obtained (eventually)?
2. Is it possible to (eventually) obtain clearly separated clusters for versicolor and virginica?
3. Can a continuous response map be obtained such that the hues provide a clearer separation of species?
4. Does the counts plot appear to change drastically with each retraining?
5. Do neighborhood distances appear to change drastically with each retraining?

6. Do the primary cluster-defining variables for each cluster in the codes plot appear to change drastically (or not at all) with each retraining?
7. How does the label variable *Species* in Figure 4.9 help reach a conclusion in all of these explorations?

A reasonable matter to ponder is whether all of the variables (sepal length, sepal width, petal length, and petal width) need to be used in the clustering process.

8. Based on the codes plot, which of the four variables might it make most sense to exclude? Explain.
9. Remove the variable identified above, then rerun the whole process again, retrain the self-organizing map several times. Does anything appear to change? If yes, for the better or for the worse?
10. Would removing another variable help the cause? Try this out. What happens?

Going back to Figure 4.8, it may be interesting to see if either, or both of the features Show Clusters and Toroidal add any useful information to the plots constructed.

11. Check the Show Clusters box and then retrain the self-organizing map a few times. What happens? Does this feature help? Explain.
12. Uncheck the Show Clusters box and then check the Toroidal box. Then retrain the self-organizing map a few times again. What happens? Does this feature help? Explain.
13. Check both the Show Clusters and the Toroidal boxes and retrain the self-organizing map a few times. Yet again, what happens, and does this help the cause any?

Going back to Figure 4.8, try increasing the SOM size and retrain the map a few times.

14. Does increasing the SOM size make the clustering clearer? If yes, how?
15. At some point, does it appear that increasing the SOM size provides no further useful information? If yes, how and why?

4.5.4 Extensions and other ideas

Notice that having the species label helped greatly in the iris clustering example. Also, this example suggests a whole range of possible flower-related extensions. For example,

–   Wild lupines have a fairly large range and it may be of interest to see if it is possible to use cluster analysis determine which features (if any) of the flowers can be used to identify its location of origin to a reasonable degree of accuracy. To do this,
    a) Identify appropriate variables that may help in the clustering; and
    b) Determine a useful "control" (like setosa in the iris example).

Of course, just like for the iris example, data for each of the variables will need to be obtained and *Region* can be used as a labels variable.
- Analyses for other wild flowers (e. g., fireweed, yarrow, etc.) can be similarly conducted.
- Similarly, regional classifications for widespread tree species can be examined, for example, yellow spruce in the Pacific Northwest of North America.

While these ideas focus on plants, it is possible to replace plants by other living beings and pose classification questions about them.

4.6 A bit more about this app

There is one small limitation. While the number of variables used to help partition the data into clusters does not appear to affect the actual clustering process, the plots do become messier as the number of numeric variables grows. In particular, the codes plot looks quite different and not as informative if the number of variables goes beyond fourteen. As an example, use the sample data set Vehicles (not a biological data set!) with all of the variables selected (the labels variable in this case is *Class*) and then look at the codes plot. Then drop one variable at a time and see what eventually happens when fourteen variables are left. Notice that even with fourteen variables this plot is not very easy to interpret!

There is a possible work-around this limitation, which actually fits in with the exploratory nature of self-organizing maps. If the number of variables is greater than fourteen, pick any fourteen (or less) variables and perform a few retraining runs of the app. Look at the codes plot and identify obvious variables that do not appear to contribute much (if at all) to the clustering. Make a note of these, then replace them with other variables and repeat the process. In the end, it should be possible to identify those variables that contribute the most to the clustering.

There may be one more matter to pay attention to in the years to follow, after the publication date of this book. This app works with version 3.0.12 of the R package kohonen and it is expected that the consistency that exists within R over time will ensure that this will remain the case with later versions. If issues do arise or any glitches in the app become evident at any point, the authors would greatly appreciate being alerted about them.

Finally, when an app session is done with, close the self-organizing app window, and then select File → Close All, and then select File → Quit Session... to shut down RStudio. Finally, in the Quit R Session pop-up window select Don't Save.

To start up a new session with the self-organizing map app, open RStudio, then navigate to the ShinySomApp folder under File and click on the Main.R file, and finally click on the Run App button (see Figure 4.3).

4.7 A fish example

The purpose of this example is to illustrate the ease with which "outside" data can be uploaded into the self-organizing map app.

The *synthetic data* for this example, inspired by the article [62], were obtained from the *Kaggle* website (see Section 4.8.2). The data set, downloaded as a *csv file from this website, could be placed in a subfolder within the ShinySomApp folder named, for example, FunData. For purposes of this example, rename the file as FishData.csv. Then, as mentioned in Section 4.4.2, it is useful to open the file and make sure the data within the file have the appropriate format and contain no rows with missing entries.

Briefly, the downloaded data set contains four columns, three of which correspond to the numeric variables *length*, *weight*, and *w_l_ratio*. The fourth is the labels variable, *species*, which contains character strings identifying each fish species by its scientific name. Again, for purposes of this example, rename the variables as *Species*, *Length*, *Weight*, and *WLRatio*. It will be noticed that the species names contain two words with a space between, and are a little long. One could choose to shorten these names. However, again for purposes of this example, leave the species names as they are, then save and close the *.csv file.

Now start up RStudio and run the app as previously described. Then click on the Import Data tab and then on the Browse button below Enter CSV file. When this is done, navigate to the previously created FunData subfolder and select and open FishData.csv. The file name will appear next to the Browse button. Finally, click on the Upload Data button (next to the Use Sample Data button) and notice it turns green. The uploading process has been completed.

To begin the clustering process, click on the Visualize Data tab, select the variables *Length*, *Weight*, and *WLRatio*, change the SOM size to 15 (just to be different), and uncheck the Toroidal box. Now scroll down and take a look at the *class* mapping plot. Notice that the long species names run over the plot—this is a good reason to have short or abbreviated labels.

On looking at the *colorless* mapping plot with species labels it will be evident over multiple retrainings that there are very few cases of cluster overlaps in the data. The majority of overlaps that occur are for the *Otolithoides biauritus* and *Setipinna taty* species. It is quite possible that the reason for these "tidy separations" lies in the synthetic nature of the data. The *continuous* mapping plot and the *codes* plots also provide some nicely organized visual representations that very clearly isolate *Sillaginopsis panijus*.

Note:
These data were downloaded on October 26, 2024, and if the synthetic data undergo any changes after this date, these observations may not remain accurate.

In a similar manner, any data set suitable for self-organizing maps may be uploaded and analyzed—the key tips being:

- The data set contains three or more meaningful numeric variables, and there are no rows containing missing data;
- Variable names are short but meaningful and a suitable labels variable is included; and
- Entries for label variables are short character strings without spaces.

The challenge, very often, is being able to find freely available data sets with these features. However, it may be possible to adapt existing data sets so that the self-organizing map app can be applied to them.

For example, suppose there are analogous (but separate) data sets for polar bears (or brown bears) from different countries within their range. Meaning, they all contain the same or equivalent measurements obtained from polar bears (or brown bears) sampled in the various regions. Then, with some appropriate adjustments and using region as the labels variable, the various data sets could be combined into a single data set to determine if the bear characteristics measured lend themselves to a clustering of the bears according to region.

4.8 Selected resources

The focus of this chapter has been on using self-organizing maps to introduce clustering and cluster analysis through basic mapping and analysis plots. This being said, there are some rabbit-holes that the interested reader may choose to go down. These may involve R (in RStudio), Shiny, finding open source data, or the broader aspects of exploratory data analysis through clustering.

4.8.1 R, RStudio, and Shiny

For those intrigued by R, an introductory user's manual (and more) can be found on the *R-Project for Statistical Computing* website at

https://cran.r-project.org/manuals.html

For those who wish to pursue further projects with R using the RStudio platform, the users guide for this can be found on the RStudio website at

https://docs.posit.co/ide/user/

The Shiny beginner's guide can be found at

https://shiny.posit.co/r/getstarted/shiny-basics/lesson1/index.html

and more on mastering Shiny can be found at

https://mastering-shiny.org/basic-app.html

It is worth noting that a reasonable level of comfort with coding in R will be very helpful for those who choose to embark on further adventures in clustering using RStudio.

4.8.2 Open source data sets

Online open source data sets can be found quite easily, but finding open data that lend themselves to self-organizing maps quickly may be a challenging task. This could be done by experimenting with a variety of keyword searches. Such searches can lead to a variety of repositories. For example:
- Government websites typically contain open data sets for educational or research purposes. For example, sites associated with the United States, Canada, the United Kingdom, Australia, New Zealand, and the European Union, to name a few, can be found.
- Many universities have data repositories for public access. For example, Boston University, University of Michigan, University of Missouri, and others can be found.
- Institutes of advanced research, such as the National Institutes of Health and the National Science Foundation have open data repositories.
- Federal and State agencies such as the U. S. Fish and Wildlife Service and the Alaska Department of Fish and Game have open data repositories. Equivalent wildlife agencies from other countries having open data repositories can also be found.
- Repositories for open data sets associated with ecology can be found either through governmental agencies, universities, or nonprofit organizations.
- The World Wildlife Fund contains an open data portal, and so does the United Nations. Open source data are available through other nonprofit agencies such as the World Health Organization and more.

The earlier mentioned *Kaggle* website

https://www.kaggle.com/

contains data sets associated with biology and ecology. The data sets available on this website are free, but registration is required.
Note: While the data used in the previous fish example is a synthetic data set, data sets on this website are not necessarily synthetic.

Finally, there is the search engine

https://datasetsearch.research.google.com/

which is maintained by Google, probably along the lines of Google Scholar.

4.8.3 Background and theory

Those interested in delving deeper into the finer details of exploring self-organizing maps with package kohonen can refer to the following that were prepared by the creators of this package.

Self- and superorganizing maps in R: The kohonen package (2007) [27]
Abstract: In this age of ever-increasing data set sizes, especially in the natural sciences, visualisation becomes more and more important. Self-organizing maps have many features that make them attractive in this respect: they do not rely on distributional assumptions, can handle huge data sets with ease, and have shown their worth in a large number of applications. In this paper, we highlight the kohonen package for R, which implements self-organizing maps as well as some extensions for supervised pattern recognition and data fusion.

Flexible self-organizing maps in kohonen 3.0 (2018) [119]
Abstract: Self-organizing maps (SOMs) are popular tools for grouping and visualizing data in many areas of science. This paper describes recent changes in package kohonen, implementing several different forms of SOMs. These changes are primarily focused on making the package more useable for large data sets. Memory consumption has decreased dramatically, amongst others, by replacing the old interface to the underlying compiled code by a new one relying on Rcpp. The batch SOM algorithm for training has been added in both sequential and parallel forms. A final important extension of the package's repertoire is the possibility to define and use data-dependent distance functions, extremely useful in cases where standard distances like the Euclidean distance are not appropriate. Several examples of possible applications are presented.

Package "kohonen" (2019) [78]
Description: This contains the relevant R documentation pages for functions and data sets contained in the package kohonen.

A selection of books with brief descriptions and further articles on self-organizing maps with abstracts and brief notes include the following works. These are listed in chronological order and go beyond the simple applications presented in this chapter. However, they are useful resources for background theory.

The self-organizing map (1998) [75]
Abstract: The self-organized map, an architecture suggested for artificial neural networks, is explained by presenting simulation experiments and practical applications. The self-organizing map has the property of effectively creating spatially organized internal representations of various features of input signals and their abstractions. One

result of this is that the self-organization process can discover semantic relationships in sentences. Brain maps, semantic maps, and early work on competitive learning are reviewed. The self-organizing map algorithm (an algorithm which order responses spatially) is reviewed, focusing on best matching cell selection and adaptation of the weight vectors. Suggestions for applying the self-organizing map algorithm, demonstrations of the ordering process, and an example of hierarchical clustering of data are presented. Fine tuning the map by learning vector quantization is addressed. The use of self-organized maps in practical speech recognition and a simulation experiment on semantic mapping are discussed.

Notes: For those interested, the author of this article is the one who developed the algorithm for self-organizing maps.

Self-organising maps: Applications in geographic information science (2008) [2]
Description: This book brings together geographical research where self-organizing maps were utilized, providing readers with a snapshot of ideas that may be adapted and used in new research projects. The book begins with an overview of the self-organizing map algorithm along with commonly used (and freely available) software (at the time of publication). Then it addresses different uses of the technique, namely clustering, data mining and cartography, from a range of application areas in the biophysical and socio-economic environments. At the time of its publication, this was apparently the only book that introduced self-organizing maps to the GIS and geography research communities. Experts from a range of countries contributed a range of case studies that involved techniques in clustering, data mining cartography.

Self-organizing maps (2012) [76]
Description: This is the fourth monograph written by the creator of self-organizing (Kohonen) maps. It contains all the technical details associated with self-organizing maps beginning with mathematical preliminaries, moving on to the justification of using neural modeling, the basic self-organizing map, and much, much more.

Applications of self-organizing maps (2012) [68]
Description: The self-organizing map, first described by the Finnish scientist, Teuvo Kohonen, can by applied to a wide range of fields. This book is about such applications. That is, how the original self-organizing map as well as variants and extensions of it can be applied in different fields. In fourteen chapters, a wide range of such applications is discussed. To name a few, these applications include the analysis of financial stability, the fault diagnosis of plants, the creation of well-composed heterogeneous teams and the application of the self-organizing map to the atmospheric sciences.

Self-organizing maps and their applications to data analysis (2019) [99]

Abstract: Self-Organizing Maps (SOMs) are a form of unsupervised neural network that are used for visualization and exploratory data analysis of high dimensional data sets. Our goal was to understand how we can use a SOM to gain insights about data sets. We do this by first understanding the initialization, training, error metrics, and convergence properties of the SOM. Next, we discuss the ways to interpret and visualize a Self-Organizing Map. Finally we used real datasets to understand what the Self-Organizing Map can tell us about labeled and unlabeled data. Based on experiments with our data sets we found that the Self-Organizing Map can tell us about the spacing and position of high dimensional clusters, help us find nonlinear patterns, and give us insight into the shape of our data.

Notes: This article provides a condensed discussion about self-organizing maps, with recommendations on when they should be used and practical advice about their use.

For those interested in pursuing further topics and methods applicable to data clustering, plenty of articles on clustering and cluster analysis in general can be found. The following selection includes papers from different time periods written by individuals involved in a range of fields of study.

The literature on cluster analysis (1978) [21]

Abstract: There has been an explosion of interest in cluster analysis since 1960. The "explosion" of this literature is documented through: (a) a rapid growth in the number of articles, which have been published using this technique; (b) the wide range of sciences interested in clustering; (c) the large and growing number of software programs for performing cluster analysis; (d) the formation of cliques of cluster analysis users; and (e) the resulting fragmentation of terminology into jargon, which restricts interdisciplinary communication. In response to the effects of this explosion, it is expected that the future literature on clustering will contain a number of attempts at consolidation. Nevertheless, the facts that cluster analysis has no scientific home, that clustering methods are not based upon a well-enunciated statistical theory and that cluster analysis is tied to the complex topic of classification means that the consolidation of this literature will be difficult.

Notes: This article, written by behavioral scientists, may be of interest for some from the historical perspective.

Methods for exploratory cluster analysis (2003) [69]

Abstract: The Self-Organizing Map is a nonlinear projection of a high-dimensional data space. It can be used as an ordered groundwork, a two-dimensional graphical display, for visualizing structures in the data. Different locations on the display correspond to different domains of the data space in an orderly fashion. The models used in the mapping are fitted to the data so that they approximate the data distribution nonlinearly but smoothly. In this paper, we introduce new methods for visualizing the cluster structures of the data on the groundwork, and for the interpretation of the structures in terms of the local metric properties of the map. In particular, it is possible to find out which variables have the largest discriminatory power between neighboring clusters. The methods are especially suitable in the exploratory phase of data analysis, or preliminary data mining, in which hypotheses on the targets of the analysis are formulated. We have used the methods for analyzing a collection of patent abstract texts. We found, for instance, a cluster of neural networks patents not distinguished by the official patent classification system.

Notes: This article may be a source for ideas on using methods from cluster analysis for exploratory data analysis and the early stages of data mining.

Clustering (2009) [124]

Brief description: The publisher describes this as the first book to take a truly comprehensive look at clustering. It begins with an introduction to cluster analysis and goes on to explore: proximity measures; hierarchical clustering; partition clustering; neural network-based clustering; kernel-based clustering; sequential data clustering; large-scale data clustering; data visualization and high-dimensional data clustering; and cluster validation. The authors assume no previous background in clustering and their generous inclusion of examples and references help make the subject matter comprehensible for readers of varying levels and backgrounds.

Overview on techniques in cluster analysis (2010) [41]

Abstract: Clustering is the unsupervised, semisupervised, and supervised classification of patterns into groups. The clustering problem has been addressed in many contexts and disciplines. Cluster analysis encompasses different methods and algorithms for grouping objects of similar kinds into respective categories. In this chapter, we describe a number of methods and algorithms for cluster analysis in a stepwise framework. The steps of a typical clustering analysis process include sequentially pattern representation, the choice of the similarity measure, the choice of the clustering algorithm, the assessment of the output, and the representation of the clusters.

Notes: This article, in the journal titled *Bioinformatics Methods in Clinical Research*, may be of interest to those focusing on applications in medical fields.

Cluster analysis: An overview (2015) [57]

Abstract: This chapter gives an overview of the basic concepts of cluster analysis, including some references to aspects not covered in this handbook. It introduces general definitions of a clustering, for example, partitions, hierarchies, and fuzzy clusterings. It distinguishes objects × variables data from dissimilarity data and the parametric and nonparametric clustering regimes. A general overview of principles for clustering data is given, comprising centroid-based clustering, hierarchical methods, spectral clustering, mixture of model and other probabilistic methods, density-based clustering, and further methods. The chapter then reviews methods for cluster validation, that is, assessing the quality of a clustering, which includes the decision about the number of clusters. It then briefly discusses variable selection, dimension reduction, and the general strategy of cluster analysis.

Notes: This chapter from the *Handbook of Cluster Analysis* provides an overview of the various methods available, and the handbook itself may be of interest to those who wish to delve deeper into applications of cluster analysis.

Cluster analysis (2024) [56]

Abstract: The next two chapters address classification issues from two varying perspectives. When considering groups of objects in a multivariate data set, two situations can arise. Given a data set containing measurements on individuals, in some cases we want to see if some natural groups or classes of individuals exist, and in other cases, we want to classify the individuals according to a set of existing groups. Cluster analysis develops tools and methods concerning the former case. That is, given data containing multivariate measurements on a large number of individuals (or objects), the objective is to build some natural subgroups or clusters of individuals. This is done by grouping individuals that are "similar" according to some appropriate criterion.

Notes: This chapter in the book *Applied Multivariate Statistical Analysis* introduces the ideas and methods involved in clustering.

4.8.4 Research and ideas

The following represent a chronological selection of articles from the literature, each presenting a possible avenue of exploration for the curious reader.

Using self-organizing maps to explore patterns in species richness and protection (2008) [61]

Abstract: The combination of species distributions with abiotic and landscape variables using Geographic Information Systems can prioritize areas for biodiversity protection by identifying areas of high richness, although the number of variables and complexity of the relationships between them can prove difficult for traditional statistical methods. The use of these methods, which commonly assume linearity and low correlation between independent variables, can obscure even strong relationships and patterns. Self-Organizing Maps (SOM) is a heuristic statistical tool based on machine learning methods that can be used to explore patterns in large, complex data sets for linear and nonlinear patterns. Here, we use SOM to visualize broad patterns in species richness by taxonomic group (birds, mammals, reptiles, and amphibians) and 78 habitat, landscape and environmental variables using data from the Gap analysis project for West Virginia, USA.

Soil and habitat variables demonstrated clear relationships with species richness; areas with high species richness occurred in areas with high soil richness. Landscape metrics were less important, although habitat diversity and evenness indices were positively related to species richness in some taxonomic groups. Current coverage of protected areas (e. g., National Forests and state parks) appeared to be insufficient to cover most of the areas of high species richness, especially for reptiles; many of the polygons with the highest richness were not covered by these areas. The identification of polygons with high richness and low protection can be used to focus conservation efforts in those areas.

Notes: This article may be of interest to individuals interested in applications of self-organizing maps to wildlife management and conservation.

The self-organizing maps: Background, theories, extensions, and applications (2008) [126]

Abstract: For many years, artificial neural networks (ANNs) have been studied and used to model information processing systems based on or inspired by biological neural structures. They not only can provide solutions with improved performance when compared with traditional problem-solving methods, but also give a deeper understanding of human cognitive abilities. Among various existing neural network architectures and learning algorithms, Kohonen's self-organizing map (SOM) [77] is one of the most popular neural network models. Developed for an associative memory model, it is an unsupervised learning algorithm with a simple structure and computational form, and is motivated by the retina-cortex mapping. Self-organization in general is a fundamental pattern recognition process, in which intrinsic inter- and intra-pattern relationships among the stimuli and responses are learned without the presence of a potentially biased or subjective external influence. The SOM can provide topologically preserved mapping from input to output spaces. Although the computational form of the SOM is very simple, numerous researchers have already examined the algorithm and many of its problems; nevertheless, research in this area goes deeper and deeper and there are still many aspects to be exploited.

In this chapter, we review the background, theories, and statistical properties of this important learning model and present recent advances from various pattern recognition aspects through a number of case studies and applications. The SOM is optimal for vector quantization. Its topographical ordering provides the mapping with enhanced fault- and noise-tolerant abilities. It is also applicable to many other applications, such as dimensionality reduction, data visualization, clustering, and classification. Various extensions of the SOM have been devised since its introduction to extend the mapping as effective solutions for a wide range of applications. Its connections with other learning paradigms and application aspects are also exploited. The chapter is intended to serve as an updated, extended tutorial, a review, as well as a reference for advanced topics in the subject.

Notes: This chapter, contained in the book *Computational Intelligence: A Compendium*, may be a useful source for those interested in more on the underlying theory of clustering.

Application of self-organizing maps for assessing soil biological quality (2008) [89]

Abstract: Agricultural ecosystems can be described by many different variables that include soil chemical, physical, and biological data. Whilst most of the chemical and physical properties variables that are relevant to soil quality are well understood, measures of soil biological properties so far have been much more difficult to use as decision support tools for monitoring soil quality. Descriptions of soil biological properties can range from single parameter variables such as microbial biomass or respiration to multiparametric data that describe biochemical profiles, measurements of enzyme activities, and molecular analyses of microbial communities. With the aim of developing practical measures of soil quality, integrative approaches are now being explored to sift out interrelationships between various types of variables. Among the different statistical tools applied, an increasing number of studies have used artificial neural networks (ANNs) to probe complex data sets. As an example of how ANN can be used, we provide an example analysis of soils from two different regions of Southeast Australia using Kohonen self-organizing maps (SOM) using data sets containing biochemical signatures of microbial communities determined by phospholipid fatty acid analysis (PLFA), genetic signatures obtained by terminal restriction fragment length polymorphisms (TRFLP), and a range of single parameter soil chemical, physical, and biological variables. The visual output of the SOM analysis provides a rapid and intuitive means to examine covariance between variables and with minimal training could be useful for assisting land managers with interpretation of multiparametric soil analyses. Further development of these tools should also help soil scientists to identify novel relationships and devise research to explore linkages between the biological, chemical, and physical properties of soils.

Notes: This article may be of use to those interested in pursuing studies in the soil sciences. It presents an exploration of using agent-based models to examine relationships between soil quality variables, along with a detailed discussion of the rationale and underlying theory.

Application of cluster analysis in agriculture—a review article (2011) [114]

Abstract: In this paper, we humbly present a review of some applications of cluster analysis in the field of agriculture and allied sciences. A few applications among them, which have been discussed here includes hierarchical agglomerative clustering approach, fuzzy clustering, hierarchical divisive clustering, and Kohonen self-organizing feature maps along with an application of each of these techniques in the field of agriculture is also presented. Data mining in agriculture is a relatively new research field and the use of cluster analysis has almost been just begun in this area. This is our strong belief that effective techniques can be carved out for solving different agricultural problems of various complexities by intelligent use of data mining and its tools such as cluster analysis.

Notes: Among other clustering methods, this article provides a review of applications of self-organizing maps to agricultural studies.

Self-organizing maps applied to ecological sciences (2011) [30]

Abstract: Ecological data are considered to be difficult to analyze because numerous biological and environmental factors are involved in a complex manner in environment organism relationships. The Self-Organizing Map (SOM) has advantages for information extraction (i. e., without prior knowledge) and the efficiency of presentation (i. e., visualization). It has been implemented broadly in ecological sciences across different hierarchical levels of life. Recent applications of the SOM, which are reviewed here, include the molecular, organism, population, community, and ecosystem scales. Further development of the SOM is discussed regarding network architecture, spatio-temporal patterning, and the presentation of model results in ecological sciences.

Notes: This article discusses applications of self-organizing maps to studies involving a range of applications, including possible future developments.

Using self-organizing maps in the visualization and analysis of forest inventory (2012) [73]

Abstract: A lot of useful data on forest condition can be gathered from the Forest Inventory (FI). Without the help of data analysis tools, human experts cannot manually interpret information in such a large data set. Conventional multivariate statistical analyses provide results that are difficult to interpret and often do not represent the information in a satisfactory way. Our goal is to identify an alternative approach that will enable fast and efficient interpretation and analysis of the FI data. Such interpretation and analysis can be performed automatically with a clustering method, but all clustering methods have some shortcomings. Therefore, our aim was also to provide information in a form suitable for fast and intuitive visualization. Kohonen's Self-Organizing Map (SOM) is an alternative approach to data visualization and analysis of large multidimensional data sets. SOM provides different possibilities and our experiments are presented with component matrices of individual stand parameters and label matrices. In forming data clusters, we experimented with hierarchical and non-hierarchical clustering methods. Our experiments showed that SOM provides useful information in a form suitable for data clustering and data visualization. This enables an efficient analysis of large FI data sets at different analysis scales. Clustering results obtained with SOM and two clustering algorithms are in accordance with ground truth. We have also considered the efficiency of SOM component matrices by visual comparison and correlation among structural parameters and by determining contributions of individual stand parameters to clustering input data. SOM application in visualization and analysis of stand structural parameters enables gathering quickly and efficiently holistic information on the current condition of forest stands and forest ecosystem development. Therefore, we recommend the application of Kohonen's SOM for visualization and analysis of FI data.

Notes: This article may be of interest to those involved in, or contemplating studies involving self-organizing maps and forestry.

Classifying habitat characteristics of wetlands using a self-organizing map (2023) [71]

Abstract: Wetlands are nutrient-rich and biodiverse ecosystems that provide habitats for various animals and plants and protect against flooding. Classification of wetlands provides information to conservation planners and resource managers for ecosystem service determination. Many ecological case studies illuminate the self-organizing map (SOM) as a robust and powerful data classification and visualization tool. In this study, we use the SOM to analyze the habitat characteristics of inland wetlands in South Korea. We surveyed the plants, benthic macroinvertebrates, and bird species inhabiting 530 nationwide wetlands for four years from 2016 to 2019. Nine environmental features, including the proportion of urban area, farmland, grassland, a forest within a 1 km buffer zone, distance from the river and nearest wetland, area, perimeter, and average slope of wetland polygons, were used to train the SOM and examine the habitat characteristics of the surveyed living components. A map size of 10×11 pixels was considered for SOM training, and the output data were classified into eight clusters. Based on the occurrence frequency of the surveyed species group, most species were distributed in all clusters, whereas some dominated in specific clusters. We believe that our study contributes significantly to the literature because it highlights the significance of the SOM approach to cluster wetlands with dependent habitats and provides ecological information to build sustainable wetland conservation policies.

Notes: This article may be of interest to those looking to apply self-organizing maps to the analysis of wetlands and their associations with living organisms.

Bibliography

[1]   Abraham, T. H. (2004). Nicolas Rashevsky's mathematical biophysics. *Journal of the History of Biology*, **37**(2), 333–385. https://doi.org/10.1023/B:HIST.0000038267.09413.0d.

[2]   Agarwal, P., Skupin, A. (2008). *Self-Organising Maps: Applications in Geographic Information Science*. Wiley.

[3]   *Agent-based modeling* (n. d.). Columbia University Mailman School of Public Health. https://www.publichealth.columbia.edu/research/population-health-methods/agent-based-modeling.

[4]   Akman, O., Bhumpelli, S., Cline, C., Hay-Jahans, C. (2024). Compartmental modeling for the neophyte: An application of Berkeley Madonna. *Spora: A Journal of Biomathematics*, **10**, 7–14. https://doi.org/10.61403/2473-5493.1089.

[5]   Akman, O., Bhumpelli, S., Hay-Jahans, C. (2024). Agent-based modeling for the neophyte: An application of Netlogo. *Spora: A Journal of Biomathematics*, **10**, 37–49. https://doi.org/10.61403/2473-5493.1095.

[6]   Akman, O., Comar, T., Gonzales, J., Hrozencik, D. (2019). Chapter 11—data clustering and self-organizing maps in biology. In: Robeva, R., Macauley, M. (Eds.), *Algebraic and Combinatorial Computational Biology* (pp. 351–374). Academic Press. https://doi.org/10.1016/E978-0-12-814066-6.00011-8.

[7]   Akman, O., Schaefer, E. (2015). An evolutionary computing approach for parameter estimation investigation of a model for cholera. *Journal of Biological Dynamics*, **9**(1), 147–153. https://doi.org/10.1080/17513758.2015.1039608.

[8]   Akman, O., Betzab-Marroquin, Z., Hay-Jahans, C., Walsh, J., Wesley, T. (2022). Clustering for the neophyte: An R Shiny app for self-organizing maps. *Spora: A Journal of Biomathematics*, **8**(1), 31–37. https://doi.org/10.30707/SPORA8.1.1647886301.847652.

[9]   Atkins, G. L. (1969). *Multicompartmental Models for Biological Systems*. Methuen, London.

[10]   Bacaër, N. (2011). *A Short History of Mathematical Population Dynamics*. Springer, London. https://doi.org/10.1007/978-0-85729-115-8.

[11]   Bacaër, N., Bacaër, N. (2011). Daniel Bernoulli, d'Alembert and the inoculation of smallpox (1760). In: *A Short History of Mathematical Population Dynamics* (pp. 21–30). https://doi.org/10.1007/978-0-85729-115-8_4.

[12]   Baktay, J., Neilan, R. M., Behun, M., McQuaid, N., Kolber, B. (2019). Modeling neural behavior and pain during bladder distention using an agent-based model of the central nucleus of the amygdala. *Spora: A Journal of Biomathematics*, **5**(1), 1–13. https://doi.org/10.30707/SPORA5.1Baktay.

[13]   Bashford, A., Chaplin, J. E. (2016). *The New Worlds of Thomas Robert Malthus: Rereading the "Principle of Population"*. Princeton University Press. https://doi.org/10.1515/9781400880959.

[14]   Bassingthwaighte, J. B., Butterworth, E., Jardine, B., Raymond, G. M. (2012). Compartmental modeling in the analysis of biological systems. In: Reisfeld, B., Mayeno, A. N. (Eds.), *Computational Toxicology* (Vol. 1, pp. 391–438). Humana Press, https://doi.org/10.1007/978-1-62703-050-2_17.

[15]   *Berkeley Madonna Version 10*. (n. d.). https://berkeley-madonna.myshopify.com/.

[16]   Berryman, A. A. (1992). The origins and evolution of predator-prey theory. *Ecology*, **73**(5), 1530–1535. https://doi.org/10.2307/1940005.

[17]   Best, A. (2023). *Introducing Mathematical Biology*. University of Sheffield Pressbooks Network Open Textbook Library, Sheffield, UK. https://open.umn.edu/opentextbooks/textbooks/1441.

[18]   Bjørnstad, O. N., Shea, K., Krywinski, M. et al. (2020). The SEIRS model for infectious disease dynamics. *Nature Methods*, **17**(6), 557–558. https://doi.org/10.1038/s41592-020-0856-2.

[19]   Blackwood, J. C., Childs, L. M. (2018). An introduction to compartmental modeling for the budding infectious disease modeler. *Letters in Biomathematics*, **5**(1), 195–221. https://lettersinbiomath.org/manuscript/index.php/lib/article/view/77.

https://doi.org/10.1515/9783111609560-005

[20]   Blashfield, R. K., Aldenderfer, M. S. (1988). The methods and problems of cluster analysis. In: *Handbook of Multivariate Experimental Psychology* (pp. 447–473). Springer. https://doi.org/10.1007/978-1-4613-0893-5_14.

[21]   Blashfield, R. K., Aldenderfer, M. S. (1978). The literature on cluster analysis. *Multivariate Behavioral Research*, **13**(3), 271–295. https://doi.org/10.1207/s15327906mbr1303_2.

[22]   Boman, B. M., Dinh, T., Decker, K., Emerick, B., Raymond, C., Schleiniger, G. (2017). Why do fibonacci numbers appear in patterns of growth in nature? A model for tissue renewal based on asymmetric cell division. *The Fibonacci Quarterly*, **55**(5), 30–41. https://doi.org/10.1080/00150517.2017.12427733.

[23]   Boyce, W. E., DiPrima, R. C., Meade, D. B. (2021). *Elementary Differential Equations and Boundary Value Problems* (12th ed.). Wiley, New York, NY.

[24]   Brauer, F., Castillo-Chavez, C., Feng, Z. (2019). *Mathematical Models in Epidemiology*. Springer, New York, NY. https://doi.org/10.1007/978-1-4939-9828-9.

[25]   Brauer, F., van den Driessche, P., Wu, J. (Eds.) (2008). *Mathematical Epidemiology*. Springer, Berlin, Heidelberg. https://doi.org/10.1007/978-3-540-78911-6.

[26]   Burges, D. N. (1975). Population dynamics: An introduction to differential equations. *International Journal of Mathematical Education in Science and Technology*, **6**(3), 265–276. https://doi.org/10.1080/0020739750060302.

[27]   Buydens, L. M., Wehrens, R. (2007). Self-and super-organizing maps in R: The kohonen package. *Journal of Statistical Software*, **21**(5), 1–19. https://doi.org/10.18637/jss.v021.i05.

[28]   Carter, N., Levin, S., Barlow, A., Grimm, V. (2015). Modeling tiger population and territory dynamics using an agent-based approach. *Ecological Modelling*, **312**, 347–362. https://doi.org/10.1016/j.ecolmodel.2015.06.008.

[29]   Chang, W., Cheng, J., Allaire, J., Sievert, C., Schloerke, B., Xie, Y., Allen, J., McPherson, J., Dipert, A., Borges, B. (2024). *Shiny: Web application framework for R* [R package version 1.9.1]. https://CRAN.R-project.org/package=shiny.

[30]   Chon, T. (2011). Self-organizing maps applied to ecological sciences. *Ecological Informatics*, **6**(1), 50–61. https://doi.org/10.1016/j.ecoinf.2010.11.002.

[31]   Claus-McGahan, E. (1998). Modeling projects in a differential equations course. *PRIMUS*, **8**(2), 137–149. https://doi.org/10.1080/10511979808965890.

[32]   Cramer, J. (2004). The early origins of the logit model. *Studies in History and Philosophy of Science Part C: Studies in History and Philosophy of Biological and Biomedical Sciences*, **35**(4), 613–626. https://doi.org/10.1016/j.shpsc.2004.09.003.

[33]   Dass, S. (2016). A compartmental model of animal behavior. *Honors Theses* (136). https://doi.org/10.32597/honors/136/.

[34]   Dayan, P., Abbott, L. F. (2005). *Theoretical Neuroscience: Computational and Mathematical Modeling of Neural Systems*. MIT Press, Cambridge, MA.

[35]   DeAngelis, D. L., Diaz, S. G. (2019). Decision-making in agent-based modeling: A current review and future prospectus. *Frontiers in Ecology and Evolution*, **6**, 237. https://doi.org/10.3389/fevo.2018.00237.

[36]   DeAngelis, D. L., Grimm, V. (2014). Individual-based models in ecology after four decades. *F1000Prime Reports*, **6**, 39. https://doi.org/10.12703/P6-39.

[37]   Diedrichs, D. R. (2019). Using harvesting models to teach modeling with differential equations. *PRIMUS*, **29**(7), 712–723. https://doi.org/10.1080/10511970.2018.1484397.

[38]   Dunja, Š. (2023, December). Agent-based modeling in the philosophy of science. In: Zalta, E. N. E. N., Nodelman, U. (Eds.), *The Stanford Encyclopedia of Philosophy*. https://plato.stanford.edu/archives/win2023/entries/agent-modeling-philscience/.

[39]   Edelstein-Keshet, L. (2005). *Mathematical Models in Biology*. Society for Industrial and Applied Mathematics, Philadelphia, PA.

[40]   Fisher, R. A. (1936). Iris data set. *UC Irvine Machine Learning Repository*. https://archive.ics.uci.edu/ml/datasets/iris.

[41]   Frades, I., Matthiesen, R. (2010). Overview on techniques in cluster analysis. In: *Bioinformatics Methods in Clinical Research* (pp. 81–107). https://doi.org/10.1007/978-1-60327-194-3_5.

[42]   Frantzen, J. (2023). *Epidemiology of Infectious Diseases: A Human View*. BRILL. https://doi.org/10.1163/9789004689657.

[43]   Fulford, G. R. (2023). Mathematical modelling using scenarios, case studies and projects in early undergraduate classes. *International Journal of Mathematical Education in Science and Technology*, **55**(2), 468–479. https://doi.org/10.1080/0020739X.2023.2244490.

[44]   Gilbert, N. (2019). *Agent-Based Models* (2nd ed.). Sage Publications.

[45]   Godfrey, K. (1983). *Compartmental Models and Their Application*. Academic Press, NY, USA.

[46]   Gooding, T. (2019). Agent-based model history and development. In: *Economics for a Fairer Society: Going Back to Basics Using Agent-Based Models* (pp. 25–36). Springer International Publishing. https://doi.org/10.1007/978-3-030-17020-2_4.

[47]   Grant, W. E., Swannack, T. M. (2011). *Ecological Modeling: A Common-Sense Approach to Theory and Practice*. John Wiley & Sons.

[48]   Grauwin, S., Goffette-Nagot, F., Jensen, P. (2012). Dynamic models of residential segregation: An analytical solution. *Journal of Public Economics*, **96**(1–2), 124–141. https://doi.org/10.1016/j.jpubeco.2011.08.011.

[49]   Grimm, V. (1999). Ten years of individual-based modelling in ecology: What have we learned and what could we learn in the future? *Ecological Modelling*, **115**(2–3), 129–148. https://doi.org/10.1016/S0304-3800(98)00188-4.

[50]   Grimm, V., Berger, U., Bastiansen, F., Eliassen, S., Ginot, V., Giske, J., Goss-Custard, J., Grand, T., Heinz, S. K., Huse, G. et al. (2006). A standard protocol for describing individual-based and agent-based models. *Ecological Modelling*, **198**(1–2), 115–126. https://doi.org/10.1016/j.ecolmodel.2006.04.023.

[51]   Grimm, V., Berger, U., DeAngelis, D. L., Polhill, J. G., Giske, J., Railsback, S. F. (2010). The ODD protocol: A review and first update. *Ecological Modelling*, **221**(23), 2760–2768. https://doi.org/10.1016/j.ecolmodel.2010.08.019.

[52]   Grimm, V., Railsback, S. F. (2005). *Individual-Based Modeling and Ecology*. Princeton University Press, Princeton, NJ. https://doi.org/10.1515/9781400850624.

[53]   Grimm, V., Berger, U., Calabrese, J. M., Cortés-Avizanda, A., Ferrer, J., Franz, M., Groeneveld, J., Hartig, F., Jakoby, O., Jovani, R. et al. (2025). Using the ODD protocol and netlogo to replicate agent-based models. *Ecological Modelling*, **501**, 110967. https://doi.org/10.1016/j.ecolmodel.2024.110967.

[54]   Grimm, V., Railsback, S. F., Vincenot, C. E., Berger, U., Gallagher, C., DeAngelis, D. L., Edmonds, B., Ge, J., Giske, J., Groeneveld, J. et al. (2020). The ODD protocol for describing agent-based and other simulation models: A second update to improve clarity, replication, and structural realism. *Journal of Artificial Societies and Social Simulation*, **23**(2), 7. https://doi.org/10.18564/jasss.4259.

[55]   Grow, A., Van Bavel, J. (Eds.) (2017). *Agent-Based Modelling in Population Studies*. Springer.

[56]   Härdle, W. K., Simar, L., Fengler, M. R. (2024). Cluster analysis. In: *Applied Multivariate Statistical Analysis* (pp. 373–406). Springer. https://doi.org/10.1007/978-3-031-63833-6_13

[57]   Hennig, C., Meila, M. (2015). Cluster analysis: An overview. In: *Handbook of Cluster Analysis* (pp. 1–19).

[58]   Hoffmann, D. S. (2015). The dawn of mathematical biology. arXiv:1511.01455.

[59]   Hooten, M. B., Johnson, D. S., Hanks, E. M., Lowry, J. H. (2010). Agent-based inference for animal movement and selection. *Journal of Agricultural, Biological, and Environmental Statistics*, **15**, 523–538. https://doi.org/10.1007/s13253-010-0038-2.

[60]   Hoppensteadt, F. (1995). Getting started in mathematical biology. *Notices of the American Mathematical Society*, **42**(9), 969–975. https://www.ams.org/journals/notices/199509/hoppensteadt.pdf.

[61]   Hopton, M. E., Mayer, A. L. (2006). Using self-organizing maps to explore patterns in species richness and protection. *Biodiversity and Conservation*, **15**, 4477–4494. https://doi.org/10.1007/s10531-005-5099-0.

[62] Hossain, M. Y., Sayed, S. R. M., Mosaddequr Rahman, M., Ali, M. M., Hossen, M. A., Elgorban, A. M.,
      Ahmed, Z. F., Ohtomi, J. (2015). Length-weight relationships of nine fish species from the Tetulia river,
      Southern Bangladesh. *Journal of Applied Ichthyology*, **31**(5), 967–969. https://doi.org/10.1111/jai.12823.

[63] Illner, R., Bohun, C. S., McCollum, S., van Roode, T. (2005). *Mathematical Modelling: A Case Studies
      Approach*. Student Mathematical Library (Vol. 27). American Mathematical Society.

[64] Iselin, S., Segin, S., Capaldi, A. (2016). An agent-based modeling approach to determine winter
      survival rates of American robins and eastern bluebirds. *Spora: A Journal of Biomathematics*, **2**(1),
      27–34. https://doi.org/10.30707/SPORA2.1Iselin.

[65] Jackson, L. J., Trebitz, A. S., Cottingham, K. L. (2000). An introduction to the practice of ecological
      modeling. *BioScience*, **50**(8), 694–706. https://doi.org/10.1641/0006-3568(2000)050[0694:AITTPO]2.0.
      CO; 2.

[66] Jackuez, J. A. (1972). *Compartmental Analysis in Biology and Medicine*. Elsevier, Amsterdam.

[67] Janssen, M. A., Ostrom, E. (2006). Empirically based, agent-based models. *Ecology and Society*, **11**(2),
      37. http://www.ecologyandsociety.org/vol11/iss2/art37/.

[68] Johnsson, M. (2012). *Applications of Self-Organizing Maps*. IntechOpen. https://doi.org/10.5772/3464.

[69] Kaski, S., Nikkilä, J., Kohonen, T. (2003). Methods for exploratory cluster analysis. In: *Intelligent
      Exploration of the Web* (pp. 136–151). https://doi.org/10.1007/978-3-7908-1772-0_9.

[70] Kendrick, P., Apenyo, T., Callender Highlander, H. (2017). Agent-based models of the green dot
      bystander violence prevention program on college campuses. *Spora: A Journal of Biomathematics*,
      3(1), 15–28. https://doi.org/10.30707/SPORA3.1Kendrick.

[71] Kim, S., Cho, K., Kim, T., Lee, C., Dhakal, T., Jang, G. (2023). Classifying habitat characteristics of
      wetlands using a self-organizing map. *Ecological Informatics*, **75**, 102048. https://doi.org/10.1016/j.
      ecoinf.2023.102048.

[72] Kingsland, S. (2015). Alfred J. Lotka and the origins of theoretical population ecology. *Proceedings of
      the National Academy of Sciences*, **112**(31), 9493–9495. https://doi.org/10.1073/pnas.1512317112.

[73] Klobucar, D., Subasic, M. (2012). Using self-organizing maps in the visualization and analysis of forest
      inventory. *iForest-Biogeosciences and Forestry*, **5**(5), 216. https://doi.org/10.3832/ifor0629-005.

[74] Knisley, J., Schmickl, T., Karsai, I. (2011). Compartmental models of migratory dynamics. *Mathematical
      Modelling of Natural Phenomena*, **6**(6), 245–259. https://doi.org/10.1051/mmnp/20116613.

[75] Kohonen, T. (1998). The self-organizing map. *Neurocomputing: An International Journal*, **21**(1–3), 1–6.
      https://doi.org/10.1016/S0925-2312(98)00030-7.

[76] Kohonen, T. (2012). *Self-Organizing Maps*. Springer, Berlin, Heidelberg.

[77] Kohonen, T. (1982). Self-organized formation of topologically correct feature maps. *Biological
      Cybernetics*, **43**(1), 59–69. https://doi.org/10.1007/BF00337288.

[78] Kruisselbrink, J., Wehrens, R. (2019). *Package 'kohonen'*. The Comprehensive R Archive Network.
      https://cran.r-project.org/web/packages/kohonen/kohonen.pdf.

[79] Lahrouz, A., Omari, L., Kiouach, D. (2011). Global analysis of a deterministic and stochastic nonlinear
      SIRS epidemic model. *Nonlinear Analysis: Modelling and Control*, **16**(1), 59–76. https://doi.org/10.
      15388/NA.16.1.14115.

[80] Levin, S. A. (Ed.) (1994). *Frontiers in Mathematical Biology*. Springer, Berlin, Heidelberg. https://doi.org/
      10.1007/978-3-642-50124-1.

[81] Mackey, M. C., Maini, P. K. (2015). What has mathematics done for biology? *Bulletin of Mathematical
      Biology*, **77**(5), 735–738. https://doi.org/10.1007/s11538-015-0065-9.

[82] *MacTutor website* (n. d.). School of Mathematics and Statistics, University of Saint Andrews, Scotland.
      https://mathshistory.st-andrews.ac.uk/.

[83] Malthus, T. R. (1986). An essay on the principle of population (1798). In: *The Works of Thomas Robert
      Malthus* (Vol. 1, pp. 1–139). London, Pickering & Chatto Publishers.

[84] Marcoline, F. V., Furth, J., Nayak, S., Grabe, M., Macey, R. I. (2022). Berkeley Madonna Version 10—a
      simulation package for solving mathematical models. *CPT: Pharmacometrics & Systems Pharmacology*,
      11(3), 290–301. https://doi.org/10.1002/psp4.12757.

[85]   Matheson, T. S., Satterthwaite, B., Callender Highlander, H. (2017). Modeling the spread of the Zika virus at the 2016 Olympics. *Spora: A Journal of Biomathematics*, **3**(1), 29–44. https://doi.org/10.30707/SPORA3.1Matheson.

[86]   Matthews, R. B., Gilbert, N. G., Roach, A., Polhill, J. G., Gotts, N. M. (2007). Agent-based land-use models: A review of applications. *Landscape Ecology*, **22**, 1447–1459. https://doi.org/10.1007/s10980-007-9135-1.

[87]   McCarthy, C. (2017). Agent based model of salmon migration in the snake river. *Whitman College*. https://www.whitman.edu/Documents/Academics/Mathematics/2017/McCarthy.pdf (accessed: May 2024).

[88]   McLane, A. J., Semeniuk, C., McDermid, G. J., Marceau, D. J. (2011). The role of agent-based models in wildlife ecology and management. *Ecological Modelling*, **222**(8), 1544–1556. https://doi.org/10.1016/j.ecolmodel.2011.01.020.

[89]   Mele, P. M., Crowley, D. E. (2008). Application of self-organizing maps for assessing soil biological quality. *Agriculture, Ecosystems & Environment*, **126**(3–4), 139–152. https://doi.org/10.1016/j.agee.2007.12.008.

[90]   Millspaugh, J., Thompson, F. R. (2011). *Models for Planning Wildlife Conservation in Large Landscapes*. Academic Press.

[91]   Mulholland, R. J., Keener, M. S. (1974). Analysis of linear compartment models for ecosystems. *Journal of Theoretical Biology*, **44**(1), 105–116. https://doi.org/10.1016/S0022-5193(74)80031-7.

[92]   Müller, J., Kuttler, C. (2015). *Methods and Models in Mathematical Biology: Deterministic and Stochastic Approaches*. Springer, Berlin, Heidelberg. https://doi.org/10.1007/978-3-642-27251-6.

[93]   Murphy, K. J., Ciuti, S., Kane, A. (2020). An introduction to agent-based models as an accessible surrogate to field-based research and teaching. *Ecology and Evolution*, **10**(22), 12482–12498. https://doi.org/10.1002/ece3.6848.

[94]   Murray, J. D. (2007). *Mathematical Biology: I. An Introduction*. Interdisciplinary Applied Mathematics (Vol. 17). Springer, New York.

[95]   Ohajunwa, C., Seshaiyer, P. (2021). Mathematical modeling, analysis, and simulation of the COVID-19 pandemic with behavioral patterns and group mixing. *Spora: A Journal of Biomathematics*, **6**, 46–60. https://doi.org/10.30707/SPORA7.1.1647885542.955876.

[96]   Owen-Smith, N. (2009). *Introduction to Modeling in Wildlife and Resource Conservation*. John Wiley & Sons.

[97]   Pavé, A. (2012). *Modeling of Living Systems: From Cell to Ecosystem*. John Wiley & Sons, Hoboken, NJ.

[98]   Pletser, V. (2017). Fibonacci numbers and the golden ratio in biology, physics, astrophysics, chemistry and technology: A non-exhaustive review. arXiv:1801.01369.

[99]   Ponmalai, R., Kamath, C. (2019). *Self-organizing maps and their applications to data analysis* (*llnl-tr-791165*) (Tech. Rep.). Lawrence Livermore National Laboratory, Livermore, CA. https://doi.org/10.2172/1566795.

[100] R Core Team (2024). *R: A Language and Environment for Statistical Computing*. R Foundation for Statistical Computing, Vienna, Austria. https://www.R-project.org/.

[101] Railsback, S. F., Grimm, V. (2019). *Agent-Based and Individual-Based Modeling: A Practical Introduction* (2nd ed.). Princeton University Press, Princeton, NJ.

[102] Reed, M. C. (2004). Why is mathematical biology so hard? *Notices of the American Mathematical Society*, **51**(3), 338–342. https://www.ams.org/notices/200403/comm-reed.pdf.

[103] Retzlaff, C. O., Ziefle, M., Calero Valdez, A. (2021). The history of agent-based modeling in the social sciences. In: Duffy, V. G. (Ed.), *Digital Human Modeling and Applications in Health, Safety, Ergonomics and Risk Management. Human Body, Motion and Behavior* (pp. 304–319). Springer International Publishing. https://doi.org/10.1007/978-3-030-77817-0_22.

[104] RStudio Team (2024). *Rstudio: Integrated development environment for R*. Posit Software, PBC. Boston, MA. http://www.posit.co/.

[105] Sandberg, I. (1978). On the mathematical foundations of compartmental analysis in biology, medicine, and ecology. *IEEE Transactions on Circuits and Systems*, **25**(5), 273–279. https://doi.org/10.1109/TCS.1978.1084473.

[106] Schelling, T. C. (1971). Dynamic models of segregation. *The Journal of Mathematical Sociology*, **1**(2), 143–186. https://doi.org/10.1080/0022250X.1971.9989794.

[107] Semeniuk, C. A., Musiani, M., Hebblewhite, S., Grindal, M., Marceau, D. J. (2012). Incorporating behavioral–ecological strategies in pattern-oriented modeling of caribou habitat use in a highly industrialized landscape. *Ecological Modelling*, **243**, 18–32. https://doi.org/10.1016/j.ecolmodel.2012.06.004.

[108] Shiflet, A. B., Shiflet, G. W. (2014). *Introduction to Computational Science: Modeling and Simulation for the Sciences* (2nd ed.). Princeton University Press.

[109] Singley, A., Callender Highlander, H. (2020). A mathematical model for the effect of social distancing on the spread of COVID-19. *Spora: A Journal of Biomathematics*, **6**, 40–51. https://doi.org/10.30707/SPORA6.1/ARDA7202.

[110] Smith, D., Moore, L. (2004). The SIR model for spread of disease—the differential equation model. *Convergence*.

[111] Stillman, R. A., Railsback, S. F., Giske, J., Berger, U. T. A. (2015). Making predictions in a changing world: The benefits of individual-based ecology. *BioScience*, **65**(2), 140–150. https://doi.org/10.1093/biosci/biu192.

[112] Talwar, S., Yust, A. E. (2023). An investigation of mitigation measures on the spread of COVID-19 in a college classroom using agent-based modeling. *Spora: A Journal of Biomathematics*, **9**(1), 60–68. https://doi.org/10.61403/2473-5493.1086.

[113] Tang, W., Bennett, D. A. (2010). Agent-based modeling of animal movement. *Geography Compass*, **4**(7), 682–700. https://doi.org/10.1111/j.1749-8198.2010.00337.x.

[114] Tiwari, M., Misra, B. (2011). Application of cluster analysis in agriculture—a review article. *International Journal of Computer Applications*, **36**(4), 43–47. https://www.ijcaonline.org/archives/volume36/number4/4478-6298/.

[115] van den Driessche, P. (2017). Reproduction numbers of infectious disease models. *Infectious Disease Modelling*, **2**(3), 288–303. https://doi.org/10.1016/j.idm.2017.06.002.

[116] van der Hoff, Q. (2017). Interdisciplinary education—a predator–prey model for developing a skill set in mathematics, biology and technology. *International Journal of Mathematical Education in Science and Technology*, **48**(6), 928–938. https://doi.org/10.1080/0020739X.2017.1285061.

[117] Vincenot, C. E. (2014). How new concepts become universal scientific approaches: Insights from citation network analysis of agent-based complex systems science. *Proceedings of the Royal Society B*, **285**(1874), 20172360. https://doi.org/10.1098/rspb.2017.2360.

[118] Wang, X., Zanette, L., Zou, X. (2016). Modelling the fear effect in predator-prey interactions. *Journal of Mathematical Biology*, **73**, 1179–1204. https://doi.org/10.1007/s00285-016-0989-1.

[119] Wehrens, R., Kruisselbrink, J. (2018). Flexible self-organizing maps in kohonen 3.0. *Journal of Statistical Software*, **87**(7), 1–18. https://doi.org/10.18637/jss.v087.i07.

[120] Wilensky, U. (1997). Netlogo segregation model. *Center for Connected Learning Computer-Based Modeling, Northwestern University*. http://ccl.northwestern.edu/netlogo/models/Segregation.

[121] Wilensky, U. (1999). Netlogo. *Center for Connected Learning Computer-Based Modeling, Northwestern University*. https://ccl.northwestern.edu/netlogo/.

[122] Wilensky, U., Rand, W. (2015). *An Introduction to Agent-based Modeling: Modeling Natural, Social, and Engineered Complex Systems with NetLogo*. MIT Press.

[123] Wilmink, F. W., Uytterschaut, H. T. (1984). Cluster analysis, history, theory and applications. In: *Multivariate Statistical Methods in Physical Anthropology: A Review of Recent Advances and Current Developments* (pp. 135–175). Springer. https://doi.org/10.1007/978-94-009-6357-3_11.

[124] Xu, R., Wunsch, D. (2009). *Clustering*. Wiley-IEEE Press. https://doi.org/10.1002/9780470382776.

[125] Yang, C., Wilensky, U. (2011). Netlogo epidem basic model. *Center for Connected Learning Computer-Based Modeling, Northwestern University*. http://ccl.northwestern.edu/netlogo/models/epiDEMBasic.

[126] Yin, H. (2008). The self-organizing maps: Background, theories, extensions and applications. In: *Computational Intelligence: A Compendium* (pp. 715–762). Springer. https://doi.org/10.1007/978-3-540-78293-3_17.

[127] Zhang, B., DeAngelis, D. L. (2020). An overview of agent-based models in plant biology and ecology. *Annals of Botany*, **126**(4), 539–557. https://doi.org/10.1093/aob/mcaa043.

Index

age-structured population models 34, 87
agent-based models 46
– ABM, agent-based models/modeling 46
– about 46
– agents and more 46
– IBM, individual-based model 46
– modular coding *see* NetLogo, coding tips
– ODD protocol 47
– predator–prey *see* predator–prey agent-based
    model
– SIR *see* SIR agent-based model
Alfred Lotka 1
applications of biomathematics, a selection of
    topics 6
artificial neural networks 106

base R 109
basic reproduction number, $\mathcal{R}_0$ 22
Berkeley Madonna 8
– about 8
– arc 10
– Axis Settings 10, 19
– Choose Variables 10
– Define Sliders 10, 16
– defining and inserting sliders 19
– defining/quantifying flows 15
– Detach Sliders 10
– downloading and installing 9
– entering global parameter values 14
– File menu 9
– flow 10
– flowchart tools 9
– forum 35
– global 10
– Graph 9
– graphics tools 16, 27
– Initial Conditions 27
– inserting arcs 15
– inserting flows 13
– inserting reservoirs 13
– New Flowchart 9
– New Flowchart Document 9
– New Graph 10
– Overlay Plots 27
– Parameters 9, 10
– Readout 16, 27
– renaming reservoirs 13

– reservoir 10
– setting initial conditions 14
– Show Sliders 10
– tutorials 35
– user guide 35
– users group 35

classical multidimensional scaling *see* Shiny SOM
    app plots
cluster, clustering, cluster analysis 106
compartmental models 8
– compartment 9, 10
– instantaneous mixing 12, 25
– predator–prey 25
– SIR 12
– state variable 12
– system in equilibrium 12
competitive learning algorithm *see* self-organizing
    maps

Daniel Bernoulli 1
density-dependent population growth 34

Fibonacci sequence 1
Fisher's iris data 115

golden ratio 1

IBM, individual-based model *see* agent-based
    models
infectious disease models
– basic reproduction number, $\mathcal{R}_0$ 22
– disease transmission rate, c 12
– endemic disease 22
– epidemic 22
– extensions of the SIR model 23
– infected individuals, I 12
– MSEIR 24, 71
– MSEIRS 24, 71
– MSIR 24, 70
– recovered individuals, R 12
– recovery rate from infection, β 12
– SEIR 24, 70
– SEIS 23, 70
– SIR 12, 53
– SIR agent-based model 53
– SIR compartmental model 12

https://doi.org/10.1515/9783111609560-006

– SIRD 23, 70
– SIRS 23, 69
– SIRV 23, 70
– SIRVD 24, 70
– SIS 23, 70
– susceptible individuals, S 12
Intercollegiate Biomathematics Alliance GitHub 109

Kohonen map 106

learning rate function *see* self-organizing maps
Leonardo of Pisa 1
Lotka–Volterra equations 32

MacTutor bibliographies 1
mapping plots 117

NetLogo 46, 48
– about 48
– agent attributes 49
– agent-based modeling environment 49
– buttons—inserting, editing, resizing, and moving
     60
– canvas—settings, resizing, and moving 50
– Code 52
– code and coding 49
– coding dictionary 49
– coding tips 71
– community and resources 50
– documentation/comment lines 52
– getting started guide 49
– Info 51
– Interface 50
– model library 49
– monitors—inserting, editing, resizing, and
     moving 62
– opening screen 51
– patches 49
– plots—inserting, editing, resizing, and moving 62
– procedure 52
– Settings 50
– sliders—inserting, editing, resizing, and moving
     56
– switch—inserting, editing, moving, and resizing
     79
– ticks 51
– turtle graphics 49
– turtles 49
– turtle(s) as breeds 57

– user manual 49
– view updates 51
– world wrap 51
NetLogo syntax
– != 79
– ;, for comment lines 53
– ask 59, 61
– breed 56, 57, 76
– color 59
– count 59
– end 52
– fd 79
– globals 56, 57
– if 59
– let 79
– lt 79
– nobody 79
– not 79
– one-of 79
– patches 61, 79
– patches-own 77, 79
– pcolor 79
– pen-down 79
– pen-up 79
– plotxy 85
– random 79
– random-float 59
– random-normal 59
– random-xcor 59
– random-ycor 59
– rt 79
– set 59
– setxy 59
– shape 59
– size 59
– tick 61
– to 52
– true and false 56
– turtles 56, 76, 79
– turtles-own 56, 57, 77, 79
– with 59
neuron *see* self-organizing maps
Nicolas Rashevsky 1
non-dimensional variables 12

Pierre Verhulst 1
population dynamics
– age-dependent models 34, 87
– competing species models 34, 88

– density-dependent models 34, 87
– predator–prey models 25, 73
predator–prey agent-based model 73
– agent attributes 77
– agent/turtles 76
– assumptions and more 73
– canvas initialization 78
– completed interface window 83
– flowchart 75
– global variables and parameters 74, 76
– initialization procedures 77
– patch attributes 77
– plot settings, counts vs. time-steps 80
– plot settings, predator vs. prey 85
– predator movement tracker 79
– sample simulation 84
– simulation procedures 79–81
– sliders 74, 76
– variables and parameter definitions 76, 77
predator–prey compartmental model 25, 31
– δ, prey death rate per predator–prey interaction
   25
– γ, predator birth rate per predator–prey
   interaction 25
– arcs 26
– assumptions 25
– Axis Settings 27
– b, prey birth rate 25
– d, predator death rate 25
– equilibrium point 32
– flowchart 28
– flows 26
– overlaying curves for solution under different
   initial conditions 31
– parameters 25
– predator growth due to predator–prey
   interactions 25
– prey deaths due to predator–prey interactions 25
– sliders 30
– solution curves for population against time 29
– solution curves for predator against prey
   population sizes 31
– stop-time 27

R 106
RStudio 109
– Console 110
– finding the Shiny app folder 110
– Global Environment 110

– Home folder 110
– installing Shiny required packages 112
– opening window with ShinySomApp folder
   opened 111

self-organizing maps 106
– activation 108
– best matching unit 107
– competitive learning 108
– competitive learning algorithm 107
– initialization 107
– input nodes 106
– learning rate function 108
– neuron 107
– output nodes 106
– Shiny app see SOM app
– synapse 107
– training 107
– unsupervised classification algorithm 106
– visualization 108
– winning weight vector 107
Shiny 106
Shiny SOM app
– analysis plots 117
– Import Data window 114
– Introduction window 114, 115
– mapping plots 117
– retraining a map 116
– starting up the app 113
– tabs 110, 115
Shiny SOM app plots
– class representation plot 119
– classical multidimensional scaling 120
– codes plot 123
– colorless mapping plot 117, 118
– continuous response map 120
– counts plot 121
– neighborhood distance plot 122
SIR agent-based model
– agent attributes 56, 57
– agents/turtles 56
– assumptions and more 53
– canvas settings 55
– count monitors 62
– flowchart 54
– Global variables and parameters 55
– initialization procedures 57
– plots 62
– running and stopping simulations 61, 63

– simulation procedures 59
SIR compartmental model 12, 20
– arcs 14
– assumptions 12
– assumptions and more 53
– basic reproduction number, $\mathcal{R}_0$ 22
– defining flows 14
– disease transmission rate, a 12
– flowchart 17
– flows 13
– globals 14
– infection rate, a 13
– initial conditions 13
– Instantaneous mixing 12
– parameters 14
– recovery rate, β 13
– recovery rate from infection, β 12
– reservoirs 13
– sliders for parameters 16
– solution 18
– solution curves 16, 19
– solve the system, Run 16

– stop-time 18
– susceptible-infected interactions 13
– time-step 18
sliders
– in Berkeley Madonna *see* Berkeley Madonna
– in NetLogo *see* NetLogo
SOM *see* self-organizing maps
supervised classification algorithms 106
synapse *see* self-organizing maps
system of first order differential equations 8
– coupled equations 9
– nonlinear 9
– predator–prey model 31
– SIR model 20

Thomas Malthus 1
timeline of selected developments in
    biomathematics 1

unsupervised classification algorithm *see*
    self-organizing maps

Vito Volterra 1